THE OSKAR KLEIN MEMORIAL LECTURES

THE OSKAR KLEIN MEMORIAL LECTURES

Vol. 1: Lectures by C. N. Yang and S. Weinberg
with translated reprints by O. Klein

THE OSKAR KLEIN MEMORIAL LECTURES

Vol. 2: Lectures by Hans A. Bethe and Alan H. Guth
with translated reprints by Oskar Klein

Editor

Gösta Ekspong

Fysikum, University of Stockholm
Sweden

World Scientific
Singapore • New Jersey • London • Hong Kong

Published by

World Scientific Publishing Co. Pte. Ltd.
P O Box 128, Farrer Road, Singapore 9128
USA office: Suite 1B, 1060 Main Street, River Edge, NJ 07661
UK office: 73 Lynton Mead, Totteridge, London N20 8DH

THE OSKAR KLEIN MEMORIAL LECTURES
Vol. 2: Lectures by Hans A. Bethe and Alan H. Guth
with translated reprints by Oskar Klein

ISBN 981-02-1450-2

Printed in Singapore.

CONTENTS

PREFACE

The Oskar Klein Memorial Lecture and Seminar in 1990 were given by Hans A. Bethe, professor at Cornell University, Ithaca, USA, and Nobel Prize winner in Physics in 1967. The 1991 speaker was Alan H. Guth, professor at Massachusetts Institute of Technology, Cambridge, USA.

Bethe's lecture in the series of Oskar Klein Memorial Lectures at Stockholm University was given on October 9, 1990, and Guth's lecture on June 4, 1991. These two lectures, intended for both students and staff, are presented in this volume. The invited speakers also gave scientific seminars of a somewhat more advanced nature. These are also included here. Two of Oskar Klein's papers are presented in English translation for the first time.

Neutrinos are at the heart of Hans A. Bethe's contributions. In the lecture he deals with the problems to theoretically account for the observations of neutrinos from the sun. He concludes that the solar model cannot be blamed for the discrepancies between the observed flux of neutrinos and the predictions of the standard solar model. His preferred explanation is one based on neutrino oscillations between neutrinos of different flavours.

The theme of Bethe's scientific seminar was the theory of supernovae, and for anyone interested in gaining an understanding of the processes going on in the collapse and explosion of a large star this lecture is to be recommended. The main problem for the theory has been to reproduce the outgoing shock wave and the ejection of matter. According to Bethe we now qualitatively understand that "the supernova is powered by the absorption of neutrinos".

The invited speaker for 1991 was Alan H. Guth, originator of the inflation scenario for the big bang. His lecture and scientific seminar dealt with various aspects of the origin of the universe. In the written version of his lecture, "The Big Bang and Cosmic Inflation", he has included the latest observations from the COBE satellite and their theoretical interpretation. The lecture is a gold-mine for anyone willing to grasp the essentials of these ideas despite not being an expert. "Inflation allows the entire universe to develop from almost nothing" would be hard to understand without Guth's expert guidance. Towards the end he elegantly points out that the driving force behind the big bang expansion — something Oskar Klein devoted most of his late efforts to finding — is coming directly from the Klein–Gordon equation. The seminar dealt with the eventual possibility of creating a new universe in the laboratory. A quote from the seminar reads: "... the false vacuum bubble appears from the outside to be a black hole. From the inside, however, it appears to be an inflation region of false vacuum with new space being created as the region expands. The region completely disconnects from the original spacetime, forming a new, isolated closed universe."

In this volume, two papers by Klein are included in English translation from the original German. Both papers have to do with early applications of Dirac's theory for the electron. The modern reader will be surpised to see how Klein and Nishina in 1928 derived their famous formula for Compton scattering. They treated the electromagnetic field as a classical wave and the electron as a Dirac field. No particles, only waves, appeared in this so-called semiclassical treatment. They got the kinematics right and they got the angular distribution right for the scattered radiation. The theory with light as particles, first advocated in 1992 by Compton himself, up to 1928 never succeeded in getting the differential cross section right; only the kinematical relations came out in agreement with measurements.

In the historically oriented paper "The Klein–Nishina Formula", an account is given of the collaboration between a Swedish theoretician and a Japanese experimentalist. New studies of the Nobel archives shed light upon the prevailing view on the struggle between a corpuscular theory for light and the wave theory. In those days physicists were predisposed either to accept light as waves, rejecting light as particles, or to accept the reverse. One new unexpected finding is that Compton was not awarded the 1927 Nobel Prize in Physics for his discovery of the particle nature of light.

The second paper by Klein deals with the surprising results obtained when he calculated the reflection of Dirac electrons at a steep potential wall. The paper contains the "Klein paradox", namely that under certain conditions more electrons are reflected than are impinging upon the wall. The negative energy states, which Klein avoided by his method to solve the Compton scattering problem, but which were important for the reflection problem, were in those days considered unphysical. Klein's paper on wall reflection of electrons was considered a severe blow to the Dirac theory. Today there is no more any paradox. The reason for the phenomenon is the creation of electron–positron pairs in the strong electric field at the wall surface.

Some autobiographic material, which has been translated from the Swedish and included here, goes beyond what was contained in *The Oskar Klein Memorial Lectures*, Vol. 1 (ed. Gösta Ekspong, World Scientific, Singapore, 1991). One may here get acquainted with Klein's interest in cosmology and his attempts to find the driving force behind the expanding system of galaxies — the meta-galaxy in his terminology. The idea that there may be several isolated expanding meta-galaxies was not foreign to him. The reader of Guth's contributions will find similarities.

Stockholm, Sweden; March 1993
Gösta Ekspong

THE OSKAR KLEIN MEMORIAL LECTURES

Oskar Klein (1894–1977)
Professor in Theoretical Physics, 1931–1962
Oil painting by Anna Klein

THEORY OF NEUTRINOS FROM THE SUN

HANS A. BETHE
Newman Laboratory of Nuclear Studies
Cornell University
Ithaca, New York 14853, USA

1. Introduction

I knew Oskar Klein. He introduced me in 1967 when I received the Nobel Prize and also I had a very nice summer excursion with him a couple of years later. I am happy to remember him.

My subject today is neutrinos from the sun. To start with I will show you the sum of the reactions that take place in the sun which I suppose are familiar to many of you. The main reaction which keeps the sun shining, which it did yesterday, is the combination of two protons to form a deuteron with the emission of a positive electron and a neutrino.

$$\mathrm{H} + \mathrm{H} \rightarrow \mathrm{D} + e^+ + \nu\,, \quad (1)$$

$$\mathrm{D} + \mathrm{H} \rightarrow \mathrm{He}^3 + \gamma\,, \quad (2)$$

$$\mathrm{He}^3 + \mathrm{He}^3 \rightarrow \mathrm{He}^4 + 2\mathrm{H}\,, \quad (3a)$$

$$\text{or } \mathrm{He}^3 + \mathrm{He}^4 \rightarrow \mathrm{Be}^7 + \gamma\,. \quad (3b)$$

The deuterons then undergo lots of further reactions, only some of which are written down here, and some of the deuterons end up as beryllium-7.

$$\mathrm{Be}^7 + e^- \rightarrow \mathrm{Li}^7 + \nu \quad (4a)$$

$$\text{or } \mathrm{Be}^7 + \mathrm{H} \rightarrow \mathrm{B}^8 + \gamma\,. \quad (4b)$$

Beryllium-7 normally decays by capturing an electron which makes the familiar nucleus lithium-7 and of course a neutrino. Very rarely the beryllium-7 will capture a proton and that makes a nucleus boron-8, which of course is a very unusual nucleus having five protons and three neutrons. As quickly as it can, it gets rid of a bit of charge and transforms in the following manner:

$$B^8 \rightarrow Be^{8*} + e^+ + \nu \,. \tag{5}$$

Alternative to (1):

$$H + H + e^- \rightarrow D + \nu \,. \tag{1b}$$

So we have at least three reactions which give rise to neutrinos. The neutrino energy from the first reaction is about 0.4 MeV; from beryllium it is 0.8 MeV and from boron you have a continuous spectrum which goes all the way up to 14 MeV.

2. Detection of Solar Neutrinos

So you have very energetic neutrinos and less energetic ones. Well, it clearly would be very nice to observe these neutrinos. It is unfortunately very difficult to observe neutrinos because they have so little interaction with ordinary matter. However, it can be done. It was pointed out by J. Bahcall many years ago that it would be good to use chlorine-37 for this detection. Chlorine-37 when it captures a neutrino can give an analogue state in argon-37 which makes this transition relatively easy, but only for neutrinos of more than 6 MeV. Anyway, this is the standard detector for these neutrinos. Argon-37 has a 35-day half life. It then decays back into chlorine-37 and it does that by capturing a K-electron and the hole in the K-shell is then filled with electrons from the L-shell. You can see the L-electron which is ejected in this manner. This experiment has been done by Raymond Davis for about 20 years. He uses a tank car full of carbon chloride. Then that makes argon-37 — very little of it, a few atoms — so he uses a carrier. He puts in a mixture of helium and argon to flush out the argon which has been formed. He puts that gas into a counter in order to count the emitted Auger electrons.

3. What Is Wrong?

Well, he has done that very patiently.[1] Indeed he found counts, he found neutrinos. However, he found only about one quarter of the predicted number. So, in one respect this was very satisfactory; the neutrinos were actually observed. In another respect it presented a puzzle because he saw only a small fraction of the predicted number. What is wrong? Well, the first possibility is that the experimenter might have been wrong. But Davis is a very careful experimenter and he checked it over and over again and it remained true that he got much fewer neutrinos than expected. Second, it is possible that the theory of the sun might be wrong. It might be that the temperature at the centre of the sun could be lower than had been calculated. Bahcall and his collaborators studied that and he went through with them very carefully and found that there was no mistake, probably. Then finally the nuclear physics experiments might have been wrong. Maybe the cross section for making boron-8 was much smaller than had been assumed. Fowler's group at Caltech went through that with great care and even I looked at it a little and we all came to the conclusion that it was all right and that there were no mistakes there.

This investigation was summarized by Bahcall and others in a very big paper in *Reviews of Modern Physics* in 1982 and you get the feeling that everything is all right about the theory.[2,3] So still we have this discrepancy. They invented something called the solar neutrino unit (SNU) — doesn't matter what that is — and the calculation says there should be $7.6(1 \pm 0.4)$ SNU and the experiment gave 2.1 ± 0.3 SNU. Some of these 7.6 come from boron-8 and so it was desirable to find a way to measure the other neutrinos. For this purpose, for a long time, it was suggested one should use gallium-71. This transforms into germanium and after many, many years of argument two experiments are now going on using gallium.

Gallium is unfortunately very, very expensive. There are two experiments. One is in Russia in the Caucasus. Baksan is the name of the laboratory and this experiment actually has been going on since the beginning of this year and even a little longer. A preliminary report was given in the spring of this year. The other experiment is a joint experiment of Germany, Italy and France in the Gran Sasso in Italy. This is also going on but so far there is no report. Obviously the Russians are richer than the West Europeans. The West Europeans have 30 tons of gallium. The Russians now have 30 tons but expect to get another 30 tons. We in America are much poorer and couldn't afford any of these experiments, but we are collaborating in the SAGE experiment in Russia.

In Table 1 the predicted rates in SNU are shown.

Table 1. Predicted rates in SNU.

	pp	pep	^{7}Be	^{8}B	Total
^{37}Cl	0	0.2	1.0	6.0	7.6
^{71}Ga	71	3	34	14	132

The total predicted for chlorine is 7.6 SNU, of which 6 come from boron-8, 1 from beryllium, 0.2 from proton–proton and the remainder come from the carbon–nitrogen reactions which I have not mentioned.

For gallium the total expected is 132 SNU. Relatively little comes from boron, a lot from the proton–proton reaction and a lot from beryllium. This is something you should keep in mind. As I mentioned already, if there were no boron-8 the rest would be 1.6 SNU, which is very close to the observed 2.1 SNU. The first and most natural hypothesis is that the boron-8 for some reason isn't there. Well, how could that be? It could be, if the sun had a lower temperature near its centre. The standard solar model says that the central temperature is a little over 15 million degrees. If you lower that to something over 13 million degrees then the boron-8 production would go down tremendously, since it goes with the 18th power of the temperature and so everything would be fine.

The trouble is that if you lower the central temperature then you also lower the rate of the proton–proton reaction on which everything depends. We know that rate very well, because after all we get so much energy, so much radiation from the sun, and that is essentially the rate of the proton–proton reaction. This is just saying in one word what I said previously, namely looking at all the possibilities Bahcall and others found it just can't be done. There is just no way how the astrophysical calculations about the sun could have been so wrong as to give this result.

4. Neutrino Conversion

Maybe the trouble is not in the sun. If it isn't the sun, it could be the neutrinos. The sun makes enough neutrinos, but something happens to the neutrinos. For quite a long time people have said, well, there are three types of neutrinos just as there are three types of leptons. Leptons, as you know, are charged particles which do not have strong interaction. We know three leptons, namely electron, μ-meson and τ-meson. The electron has a mass of about half a million eV, a μ-meson about 100 MeV and a τ-meson 1700 MeV. To each type of lepton there should belong a neutrino. In fact, we have known that since the 1960's and the 1988 Nobel Prize was given for that discovery.

There should be three types of neutrinos — electron neutrino, μ-neutrino and τ-neutrino — and their masses should go with the lepton mass, that is the mass of the τ-neutrino should be bigger than that of the μ-neutrino, which in turn is bigger than the mass of the electron neutrino. There is no experimental confirmation of that; it is just reasonable to assume that. First of all it is reasonable to assume that neutrinos, like any other particle, have some mass and by mass, of course, I mean rest mass. It is reasonable, in analogy with other particles, to assume that neutrinos in vacuum are not the same as the neutrinos which you get from nuclear reactions. These two assumptions are reasonable to make.

We know that from the K-meson for instance. If the neutrinos in vacuum are called ν_1, ν_2, ν_3, then ν_1 is almost equal to the electron neutrino but not equal. If that is so we should be able to observe in the laboratory oscillations of neutrinos, where you start from one flavour and then end up after some distance in the other flavour. This has not been observed in spite of many attempts to do so. I think at the end of this lecture you will understand why it has not been observed.

The decisive step was taken in 1985 by two Russian theorists, Mikheyev and Smirnov.[4] It is enough here to consider just two neutrino types, the e- and the μ-neutrino. Well, they say that the e-neutrino is some factor times ν_1 + some other factor times ν_2, and the factors are chosen so that the sum of the squares is one so as to normalize and the μ-neutrino is orthogonal to the electron neutrino.

$$\begin{aligned} |\nu_e\rangle &= \cos\theta|\nu_1\rangle + \sin\theta|\nu_2\rangle\,, \\ |\nu_\mu\rangle &= -\sin\theta|\nu_1\rangle + \cos\theta|\nu_2\rangle\,. \end{aligned} \tag{6}$$

Theta is known as the mixing angle. The ν_1 and ν_2, the free space neutrinos, have certain masses m_1 and m_2. If I then transform this to the representation of electron- and μ-neutrino I get a mass matrix whose important part is the second part where the difference of the mass squares comes in. In ν_1, ν_2 representation:

$$M = m^2 = \frac{1}{2}(m_1^2 + m_2^2)\begin{pmatrix} 1 & 0 \\ 0 & 1 \end{pmatrix} + \frac{1}{2}(m_2^2 - m_1^2)\begin{pmatrix} -1 & 0 \\ 0 & 1 \end{pmatrix}.$$

Transforming this to the ν_e, ν_μ representation gives

$$M = m^2 = \frac{1}{2}(m_1^2 + m_2^2)\begin{pmatrix} 1 & 0 \\ 0 & 1 \end{pmatrix} + \frac{1}{2}(m_2^2 - m_1^2)\begin{pmatrix} -\cos 2\theta & \sin 2\theta \\ \sin 2\theta & \cos 2\theta \end{pmatrix}$$

(the first line is ν_e, the second is ν_μ).

That's all well and good. Now, if you put the neutrinos in matter then the neutrinos will interact with the electrons and this interaction will come about both by the neutral weak interaction and by the charged weak interaction. The neutral weak interaction is the same for all types of neutrinos and therefore it doesn't change the relative masses, but the charged weak interaction gives a peculiar interaction for the electron neutrino:

$$\frac{G}{\sqrt{2}}\nu_e\gamma_\lambda(1+\gamma_5)\bar{e}e\gamma_\lambda(1+\gamma_5)\bar{\nu}_e = \frac{G}{\sqrt{2}}\nu_e\gamma_\lambda(1+\gamma_5)\bar{\nu}_e(e\gamma_\lambda\bar{e}+\cdots)\,, \qquad (7)$$

using the Fierz transformation. For electrons at rest only $\gamma_4 \neq 0$; for neutrinos

$$\gamma_4 = 1,\ (1+\gamma_5) = 2;\ e\gamma_4\bar{e} = N_e,\ \text{so } V = G\sqrt{2}\,N_e\,. \qquad (8)$$

The electron neutrino can exchange with an electron and that was discovered earlier by Wolfenstein,[5] which Smirnov made use of. This interaction is equivalent to giving the electron neutrino an extra mass due to the possibility of its interaction with the electron. This extra interaction is equivalent to a repulsive potential for electron neutrinos which is written down as the last equation. Now what does that mean? We know that generally in special relativity theory

$$k^2 + m^2 = (E-V)^2 \approx E^2 - 2EV\,.$$

If you have a repulsive potential then instead of energy squared you have energy minus potential squared (k is the momentum). If I neglect V squared, which is extremely small, this equation is equivalent to changing the mass of the electron neutrino, namely adding to it twice the neutrino energy times its repulsive potential,

$$m_i^2 = 2EV = 2\sqrt{2}\,GN_eE = 2\sqrt{2}\,G\frac{Y_e}{m}\rho E \equiv A\,. \qquad (9)$$

The repulsive potential we know from Wolfenstein; G is the universal Fermi interaction, next is the density of electrons which is the ordinary density times Y_e, the ratio of electrons to nucleons. I call that product A, which is an extra mass squared for the electron neutrino. Thereby the mass matrix, which I wrote down before, gets an extra term:

$$M = m^2 = \frac{1}{2}(m_1^2+m_2^2)\begin{pmatrix}1 & 0\\ 0 & 1\end{pmatrix} + \frac{1}{2}\begin{pmatrix}A-\Delta\cos 2\theta & \Delta\sin 2\theta\\ \Delta\sin 2\theta & -A+\Delta\cos 2\theta\end{pmatrix},$$

where $\Delta = m_2^2 - m_1^2$. I have written it in such a way that its trace is zero, with $(1/2)A$ for the electron neutrino and $-(1/2)A$ for the μ-neutrino.

The quantity A is proportional to the density of matter. Therefore the electron neutrino mass changes with density — with the matter density. It is useful just to plot as in Fig. 1 the masses against the density of matter. At low

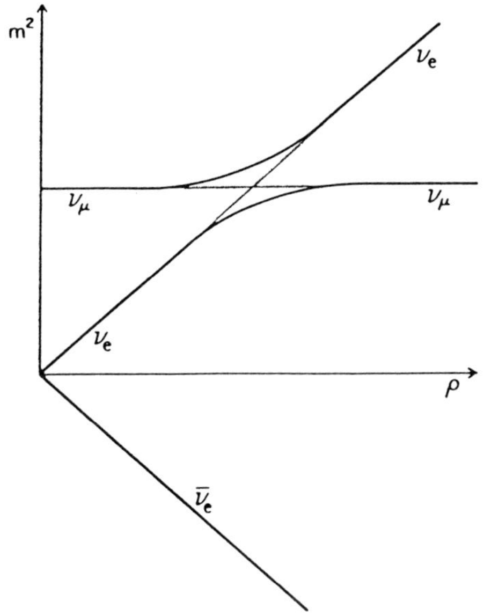

Fig. 1. The masses of two flavours of neutrinos as a function of matter density. The curves nearly cross for a particular value of the density. The mass of the electron antineutrino is also shown.

density the μ-neutrino has a higher mass than the electron neutrino. At high density the reverse is the case. I get a curve, that's the upper mass level, which at high density represents an electron neutrino, at low density a μ-neutrino. What is going to happen? A neutrino — an electron neutrino — is emitted at high density in the sun and as the neutrino goes out of the sun it slides down this curve and when it comes out of the sun it isn't an electron neutrino, it's a μ-neutrino. You can't observe μ-neutrinos of low energy, because the only way you could observe them is if they interacted with the nucleus and got converted back into a μ-meson. In order to do that it would need an energy of 100 MeV. None of the neutrinos from the sun have that much energy, so with any detector like chlorine or gallium it is impossible to observe μ-neutrinos. It will be possible for neutrinos to be formed as observable electron neutrinos at high density in the sun, but when they come out they are μ-neutrinos and cannot be observed. In principle, it would be possible that we might not observe any neutrinos from the sun at all. In fact, things are not as bad as that; we do observe some neutrinos.

Antineutrinos have an attractive potential and therefore their effective mass will go down with increasing density and so there will not be any crossing, any resonance, between μ- and electron antineutrinos. The change is of course a resonance and this resonance exists for neutrinos and not for antineutrinos. That is the theory by Mikheyev and Smirnov of 1985 and the simplest hypothesis now is again to say let us eliminate the boron-8 neutrinos. Suppose all the boron-8 neutrinos are converted into μ-neutrinos as they go through the sun and through the resonance. This is the conclusion I drew in 1986 and taking the energy of the boron neutrinos to be 10 MeV and using the formula which I had before — well, it's the resonance condition — I find that the difference of the squares of the masses of the two neutrinos has to be less than 10^{-4} $(\mathrm{eV})^2$. So assuming that the μ-neutrino mass is much bigger than the electron neutrino mass this means that the mass of the μ-neutrino is less than 10^{-2} eV compared with 500,000 eV for the electron itself. A tiny, tiny mass but not zero.

All right, well, that would be a reasonable hypothesis. You then have to assume further that the beryllium neutrinos are not converted, so the Δm^2 must be greater than 10^{-5} $(\mathrm{eV})^2$ for that purpose. In other words beryllium neutrinos would be formed below the resonance energy (Fig. 1), below the resonance density, and they would remain electron neutrinos all the way.

About the same time as I did that analytic work, Rosen and Gelb confirmed this solution by numerical calculation but they also found a second solution which also would explain the chlorine observation.[6] The second solution would have a lower difference of mass between the two neutrino types and a larger mixing angle (θ). In the second solution the neutrinos from boron-8 are only partially converted and so are those from beryllium-7. Figure 2 shows the result by Rosen and Gelb. Plotted horizontally is the energy in MeV and vertically is the probability for a neutrino of that energy to arrive on earth. And you see the stars are the solution which I found. In that solution Δm^2 is 10^{-4} $(\mathrm{eV})^2$ and boron-8 neutrinos which are high energy have very little chance to arrive on earth, but low energy neutrinos have a very good chance.

In the other solution, the Rosen–Gelb solution, the reverse is the case. How does this other solution work? This has been investigated by a number of people. The first one was Haxten at Seattle[7] and then Park at Fermi National Laboratory.[8] The best solution was given by Pizzochero,[9] a very young Italian physicist who has been working at Stony Brook for the last two years. He says, let us assume that in the sun the density is an exponential function of the distance from the centre. And that is very nearly true. The distance, R_s, over which the density decreases by a factor e is accurately known for the sun

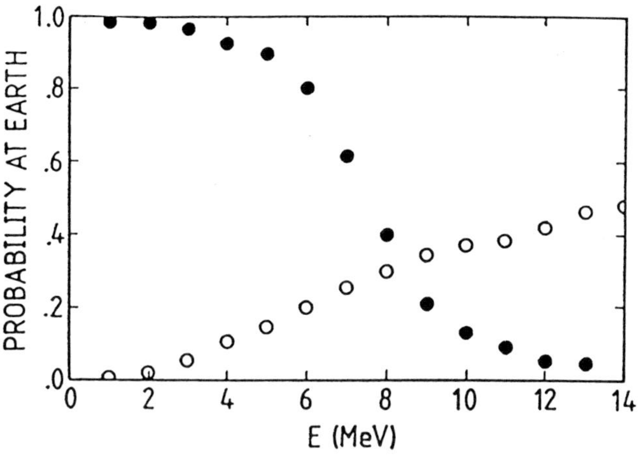

Fig. 2. Probability that electron neutrinos from the sun arrive at Earth vs. neutrino energy, E. The symbol o is for $\Delta m^2 = 1.1 \cdot 10^{-6}$ eV^2; • for $\Delta m^2 = 1.0 \cdot 10^{-4}$ eV^2 ($\sin^2 2\Theta = 10^{-1.5}$ has been assumed).

to be $R_s = 6.6 \cdot 10^9$ cm (66,000 km). With this density distribution a neutrino of a given flavour will survive the resonance, that is, remain of the same flavour — an electron neutrino remains an electron neutrino with a probability $p = e^{-C/E}$, where E is the energy of the neutrino and the constant is very simply given by $C = \pi R_s \Delta m^2 \sin^2 \theta$. That is the general theory. Unfortunately, nobody knows and quotes Pizzochero's paper but they quote the other papers, which are more complicated, give a more complicated result and are less accurate.

Most people have done numerical calculations and Fig. 3 shows one of the early and very good numerical calculations. Plotted here are the contours which would give you a certain number of SNU's in the Davis experiment. Remember Davis observed 2.1 SNU and the central curve here is precisely that. And you can have your choice. Horizontally plotted is the mixing angle and vertically the Δm^2. You can for 2.1 SNU have three different alternatives. The first solution is constant Δm^2. The second is a sloping solution where $\Delta m^2 \sin^2 \theta$ is constant, the Rosen–Gelb solution. And finally a solution, particularly found by the French group, with a large mixing angle and fairly arbitrary mass difference. I think it is unlikely to be true because everything we know about other particles. It is true that the free particle and the particle which comes out of interactions are different, but the mixing angles generally are small. I remind you of the Cabbibo angle which is the mixing angle for weak interactions. The Cabbibo angle has a $\sin^2 \theta = 0.05$, so I consider it

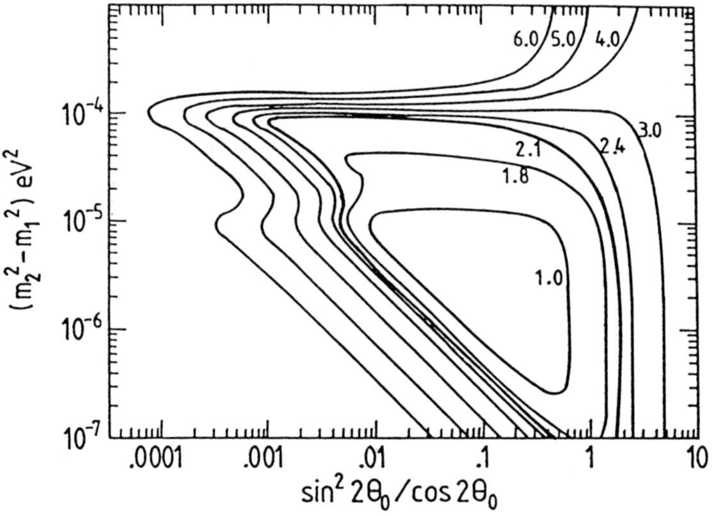

Fig. 3. Expected results for the chlorine experiment. Iso-curves in SNU in dependence of neutrino mass and mixing angle.

thereotically very unlikely that for neutrinos it should be different and that the mixing angles should be very large as this solution indicates.

From the theoretical point of view I say the alternatives are the horizontal line and the sloping line in Fig. 3. From the Davis experiments you couldn't tell whether you are on the horizontal line or on the sloping line. But then came the new experiment — real information always comes from experiments and we theorists only have to read the experimental results and try to make the best of them.

5. A New Experiment

The new experiment was done at Kamioka in Japan by a big team, most of them Japanese, but some of them American. In this laboratory one had observed the neutrinos from the supernova explosion SN1987A. They also can observe neutrinos from the sun. How do they do it? They have a big water tank about the size of this room and in this water tank — very pure water — the neutrinos collide with electrons. They project electrons and then you can measure both the energy and the direction of these electrons. And very reasonably the Kamiokande people say, let's just take those electrons which come almost from the sun, not quite because of course the electrons do not go exactly in the same direction as the neutrinos, which come almost from the sun. They take an angular distribution and subtract the background which is averaged over all directions and then they get some excess electrons which come more or less from the sun and that is what they report.[10] And there are not 50 authors but only 38. They have a minimum electron energy of 7.5 MeV and therefore they can only observe boron-8 neutrinos and they observe lots and lots of counts from which it follows that boron-8 neutrinos exist, are made, and come to the earth. So the first explanation, namely that the boron-8 is absent, is just wrong. This one experiment proved the hypothesis to be wrong. Therefore no matter what you do to the distribution of temperature in the sun you cannot explain the solar neutrino experiment. It cannot be explained by any strange assumptions about the sun. Also it cannot be explained by taking the horizontal line in Fig. 3, because in that case the boron-8 neutrinos would be converted into μ-neutrinos and be unobservable. That leaves only the slanting line.

The two experiments together, Kamiokande and Davis, show you that you have to be on the slanting line and show you also that you have to attribute the result to a conversion of neutrinos into unobservable ones. You cannot explain it by assuming that the sun doesn't behave right. The Kamiokande people put their result in terms of the ratio of what they observed to the prediction from the standard solar model. And they say that the ratio is 0.46 ± 0.06 (systematic error) $\pm$ 0.05 (statistical error). They observed for a little over 1000 days. The chlorine experiment on the other hand gave a ratio to the prediction of not 0.46 but 0.27 ± 0.04. These two numbers are different and therefore you have to explain that.

Thus the conclusions:

1. Boron-8 neutrinos do reach the earth.
2. The MSW theory (Mikheyev–Smirnov–Wolfenstein) with relatively high Δm^2 is excluded. Therefore that leaves the MSW with low Δm^2—less than 10^{-5} $(\text{eV})^2$. It is significant that the chlorine experiment gives a lower ratio to prediction than Kamiokande, which precisely is what the slanting line gives you. I showed you the prediction by Rosen and Gelb. The low energy neutrinos have less probability to remain electron neutrinos than the high energy neutrinos. That is the striking and important result.

6. A Proposed Solution

Well, seeing this, Bahcall and myself wrote a paper[11] which should be published in *Physical Review Letters.* We use the probability of escaping the resonance. According to Pizzochero, this probability is $p = e^{-C/E}$. We determined the constant C to give the observed rate in the chlorine experiments and that comes out to be $C = 10.5 \pm 5.5$ MeV. The reason why it is so uncertain is that the standard solar model doesn't give the amount of boron-8 very accurately and so we need that large uncertainty.

So this being so and taking now the beryllium neutrinos which have an energy of about 1 MeV, the survival probability ($e^{-C/E}$ with $E = 1$ MeV) is negligible or just exceedingly small. So we do not get any beryllium neutrinos to the earth. We do not get any proton–proton neutrinos to the earth. They all become μ-neutrinos. I shall modify that a little bit in a few minutes. But in first approximation that's it. Now if we take the constant C from the Davis chlorine experiment we can then calculate what the Kamiokande result should be and the answer is 0.46. Exactly equal to the measurement and it has no reason to be that exact, since I showed you that there was an experimental error of 0.06 or so. We don't think that this can be an accident and we think that this is a very good way to verify the sloping solution, the Rosen–Gelb solution.

I said already that with the probability being $e^{-C/E}$, the probability of getting any beryllium neutrinos is very, very small, $p = 5 \cdot 10^{-6}$. What about the gallium experiment? Bahcall, in particular, made propaganda for years to have the gallium experiment done. But now we say, if this is the solution then the gallium experiment won't give any answer. In fact, even in gallium, most of the counts will be due to boron-8 neutrinos. At this stage we would say that in SAGE and GALLEX we would get only 4 SNU instead of 130, about 3% of the theoretical prediction. I'll come back to that in a minute. Now using the constant C we can calculate the product

$$\Delta m^2 \sin^2 \theta = (1.0 \pm 0.5) 10^{-8} \, (\mathrm{eV})^2 \,. \tag{11}$$

That neutrino mass is not very large and therefore that's the reason why it is essentially impossible in the laboratory to observe the oscillation of neutrinos between different flavours. Remember, that Δm^2 presumably is the mass of the μ-neutrino. The mass of the electron neutrino will be much, much less still.

If I take for θ the Cabibbo angle, then $\Delta m^2 = 2 \cdot 10^{-7}$ $(\mathrm{eV})^2$, which means that the μ-neutrino will have a mass of about half a milli-electron volt. I said that this prediction was not quite correct because we must remember that from the sun to the earth the neutrinos have to go through vacuum. At the resonance in the sun the electron neutrinos, let us say coming from beryllium, are converted into μ-neutrinos, but when they come out into vacuum, I must again say that they are partly ν_1 and partly ν_2. And then the ν_1 and ν_2 will propagate in space and when they arrive at the earth they can again be observed, and when they are observed I am interested in what's the likelihood that they go back to electron neutrinos. Well, it's easy to calculate that the probability is

$$\Phi = 2\sin^2\theta\cos^2\theta\,. \tag{12}$$

The total flux of converted neutrinos, namely those which have become μ-neutrinos, is $132 - 5 = 127$, which were preserved. Therefore the predicted count of gallium is really $4.5 + 2 \cdot 127 \sin^2\theta \cos^2\theta$. We don't know the value of the mixing angle θ. That's very nice, because if they can measure accurately enough in the gallium experiment then they can determine this number and if they can do so then we shall know the mixing angle θ. We know already the product $\Delta m^1 \sin^2\theta$ and then we also know the Δm, the difference in mass.

It will be quite difficult to do the gallium experiment accurately enough, but I have talked to the American collaborators in the SAGE experiment, which is the Russian experiment, and they say, well, it's difficult but we'll try. If $\sin^2\theta$ for instance is 0.05, then instead of 4 SNU they should observe 16 SNU. That still is difficult. At the present time the SAGE experiment has given a preliminary answer. Namely they answer, we see nothing which is very nice and in agreement with our opinion and the maximum is maybe 50 SNU. It's a long way from 50 to 16 in the presence of background, but I think there is hope.

There is also an upper limit to Δm^2 which says it is certainly less than 10^{-4} (eV)^2. We are very hopeful that the gallium people actually can do as well as I would like them to do and can tell us the mixing angle. If they can then we shall know for the neutrino, the μ-neutrino probably, both the mass and the mixing angle and that would be, of course, a very important result.

Note Added in Proof

By now (May 1993) results have been published by both GALLEX[16] and by SAGE. GALLEX finds

$$83 \pm 19 \text{ (stat)} \pm 8 \text{ (syst) SNU}\,. \tag{13}$$

This is well below the prediction from the solar model, 131 ± 6 SNU. Their result is fit well by the MSW theory with

$$\Delta m = (2 \pm 0.6) \times 10^{-3} \text{ eV}\,. \tag{14}$$

[16] GALLEX collaboration, *Phys. Lett.* **B285**, 376 and 390.

The SAGE result is somewhat lower than (13), but within the given limits of error.

7. Final Comments

Two small additions. Davis has claimed that there is a correlation between his neutrino measurements and sun spots.[12] The Japanese group has, for a thousand days, when the solar activity increased 15-fold, found no such correlation. They have measured first at a very quiet sun and now at a very active sun and found no difference. Since I like there to be no correlation I believe Kamiokande. At present there are just two experiments contradicting each other. A number of Russian theorists have tried to explain the Davis result of correlation to sun spots by giving the neutrino a complicated kind of magnetic moment.[13] One important experiment in the planning stage and approved is the so-called SNO experiment in Sudbury, Ontario. Using a big tank of heavy water rather than light water, you can disintegrate the deuteron by the neutral weak current giving proton plus neutron or by the charged current giving proton plus proton and that would be very interesting.

They can also see electron recoils which give different spectra from μ-neutrinos and from electron neutrinos, so they will be able to confirm or dispute our idea that many of the neutrinos will be converted. Once you get the total number of neutrinos by these experiments you can determine how much boron-8 is actually produced in the sun and thereby you can confirm the standard solar model. And that essentially was the aim of the neutrino experiment. People wanted to see to the centre of the sun and neutrinos permit you to do that. Already now Bahcall has calculated that the standard solar model is very nicely confirmed. But only to $\pm 15\%$.

Finally some theory. There is a theory which is known as the see-saw mechanism, which was proposed independently by two groups.[14,15] That says that the mass of a neutrino of flavour i is proportional to the square of the mass of the corresponding charged particle, chosen to be a lepton or a quark:

$$m(\nu_i) = \frac{m^2(q_i)}{M_{\rm GUT}} ,$$

where $M_{\rm GUT}$ is the mass of the Grand Unified Theory.

If we believe the mass of the μ-neutrino to be $5 \cdot 10^{-4}$ eV and then choose for the charged particle not the muon but the charmed quark with mass $m(q_{\rm ch}) =$ 1.5 GeV, then the mass of Grand Unified Theory would be of the order of 10^{13} GeV, which is the sort of thing that particle theorists like. If that theory is right, then the mass of the electron neutrino would be 10^{-8} eV, so it is not very likely for you to find that by looking at the upper limit of the spectrum of electrons from the decay of hydrogen-3. The mass of the τ-neutron may be 10 eV.

I have tried to show you that the problem of the solar neutrinos can be understood. That it's done in terms of the MSW theory. That we know then accurately the product of the mass of the μ-neutrino times the sine of the mixing angle and that we can hope with some luck that the gallium experiment will tell us the mixing angle itself and thereby will tell us everything at least about the μ-neutrino. That still is not the whole story but it is a nice part of the story.

References

1. R. Davis Jr., B. T. Cleveland and J. K. Rowley, *AIP Conf. Proc. Nr* **72** (1981).
2. J. N. Bahcall, W. F. Huebner, S. W. Lubow, P. D. Parker and R. G. Ulrich, *Rev. Mod. Phys.* **54**, 767 (1982).
3. J. N. Bahcall, B. T. Cleveland, R. Davis and J. K. Rowley, *Ap. J.* **292**, 79 (1985).
4. S. P. Mikheyev and A. Y. Smirnov, *Nuovo Cimento* **C9**, 17 (1986).
5. L. Wolfenstein, *Phys. Rev.* **D17**, 2369 (1978).
6. S. P. Rosen and J. M. Gelb, *Phys. Rev.* **D34**, 969 (1986).
7. W. C. Haxton, *Phys. Rev. Lett.* **57**, 1271 (1986).
8. S. J. Parke, *Phys. Rev. Lett.* **57**, 1275 (1986).
9. P. Pizzichero, *Phys. Rev.* **D36**, 2293 (1987).
10. K. S. Hirata *et al.*, *Phys. Rev. Lett.* **63**, 16 (1989); **65**, 1297 (1990).
11. J. N. Bahcall and H. A. Bethe, *Phys. Rev. Lett.* **65**, 2233 (1990).
12. B. Cleveland *et al.*, *Proc. 25th Int. Conf. on High Energy Physics* (Singapore, 1990) (World Scientific, 1991), p. 667.
13. M. B. Voloshin, M. I. Vysotsky and L. B. Okun, *Sov. Phys. JETP* **64**, 446 (1986).
14. T. Yanagita, *Prog. Theor. Phys.* **B315**, 66 (1978).
15. M. Gell-Mann, P. Ramond and Slansky, in *Supergravity* (North-Holland, Amsterdam, 1979).

SUPERNOVA THEORY

HANS A. BETHE
Newman Laboratory of Nuclear Studies
Cornell University
Ithaca, New York 14853, USA

Supernovae are very spectacular phenomena. In a short time, a supernova produces as much energy as the sun in its entire lifetime of about 10^{10} years, namely about 10^{51} ergs (1 foe for fifty-one ergs). It is generally agreed that all elements heavier than helium came into the world from supernova explosions. There are two types of supernovae: type I originates from white dwarfs, type II from massive stars ($\geq 8M_\odot$, where $M_\odot$ is the mass of the sun) at the end of their life. Only type II will concern us here.

Nuclear reactions provide the energy of stars, converting H to He, then to C and O, then to Si, and finally to Fe or some element near it. The reactions proceed fastest near the centre of the star. Once Fe has been reached, no further energy can be extracted from the nuclei. Pressure is then provided in the core by the Fermi pressure of electrons. But there is a limit to this, the Chandrasekhar mass,

$$M_{\mathrm{Ch}} = 5.8 Y_e^2 M_\odot \,, \tag{1}$$

where Y_e is the number of electrons per nucleon. (There are some corrections to this, for various reasons.) Typically, in the core of a star late in its evolution, $Y_e = 0.45$, and after corrections, $M_{\mathrm{Ch}} = 1.3 M_\odot$. Once the mass of the iron core exceeds this limit, the core collapses due to gravitation.

The collapse is nearly adiabatic and can therefore be calculated easily. However, one change takes place: electrons are captured by nuclei because their Fermi energy exceeds the energy which has to be added to a nucleus to convert one of its protons into a neutron. In this capture, a neutrino is emitted. If this process went on indefinitely, Y_e would become very small, and so would the Chandrasekhar mass, (1). Fortunately, however, once the density exceeds about 10^{12} g cm^{-3}, neutrinos get trapped because they are scattered by the nuclei present, a process due to the weak neutral current. The trapped neutrinos cause the reverse reaction: they are captured by nuclei, and an electron is emitted. An equilibrium is reached, and the final lepton fraction (number of electrons plus neutrinos per nucleon) is about $Y_L = 0.38$, giving (after suitable corrections, including general relativity)

$$M_{\rm Ch} \simeq 0.7\ M_\odot\,. \tag{2a}$$

Neutrinos are also scattered by electrons, which decreases their energy. They can then escape trapping more easily, as shown by Bruenn and others; Y_L is decreased to about 0.36 and

$$M_{\rm Ch} \simeq 0.6\ M_\odot\,. \tag{2b}$$

The collapse is stopped when the density at the centre becomes appreciably bigger than nuclear density,

$$\rho_0 \simeq 2.5 \cdot 10^{14}\ \text{g cm}^{-3}\,. \tag{3}$$

Then nuclei merge and form continuous nuclear matter which is quite incompressible. Infall is stopped and a pressure wave moves out. This turns into a shock wave when it reaches the surface of the Chandrasekhar sphere; the inside of that sphere remains unshocked. The material outside the sphere bounces back vigorously. This is called the "prompt shock".

The prompt shock starts moving with a velocity of 2–3 × 10^9 cm/s, then gradually slows down. It dissociates the existing heavy nuclei into nucleons. This process costs a lot of energy, about 9 MeV per nucleon = $8.5 \cdot 10^{18}$ ergs per g. In addition, once that dissociation takes place, the protons eagerly capture electrons, emitting neutrinos and thus losing further energy. The shock has only limited energy, perhaps 10^{52} ergs, and so it can penetrate only a limited mass, about 10^{33} g (0.5 solar masses); but this number is quite uncertain. It is therefore important that the Chandrasekhar mass of the unshocked core be large. We found above that this mass is about 0.6 $M_\odot$ so that the shock can penetrate out to about 1.1 $M_\odot$.

Computations have shown that the shock works better if general relativity is used. This is because, in general relativity, strong gravitational effects are enhanced, and one gets stronger collapse followed by stronger rebound. In fact, Baron, Cooperstein and Kahana (1985) obtained shocks which penetrated through the iron core. It was believed that if the shock penetrated the Fe, it would go through the entire star and expel its mass in a supernova display. The reason for this belief was that in the Si + S region surrounding the core, and even more in the O region, nuclear reactions were likely to release energy rather than absorb it. This assumption was never put to the test because the shock never progressed this far in the computation, but the question has now become irrelevant.

Unfortunately, the calculations of Baron *et al.* were invalidated by Bruenn's results on the collisions between neutrinos and electrons during the infall. As I mentioned earlier, these collisions interfere with the trapping of neutrinos, and substantially reduce the mass of the unshocked core. With this reduced mass, the shock gets stuck much earlier. Baron and Cooperstein (1990) found that even with a soft equation of state for nuclear matter, and with general relativity, the prompt shock will fail for Fe core masses above 1.1 $M_\odot$, and even for this low mass it will work only if some other conditions are also favourable. On the other hand, the calculations of pre-supernova evolution now appear to be rather firm. In spite of some corrections in the physics, the mass of the Fe core remains the same as it has been since 1985, viz. about 1.3 $M_\odot$. Thus it seems rather definite that the prompt shock will not work.

James Wilson in 1982 discovered an alternative mechanism. He left his computer running overnight, and found that the shock, previously stalled, revived. His analysis showed that this was due to the absorption of neutrinos at intermediate distances, around 100 km. These neutrinos are emitted by the core, which will ultimately become a neutron star. At this stage, the core contains the energy set free by the gravitational collapse, typically $3 \cdot 10^{53}$ ergs or 300 foes. This energy is now mostly in the form of heat, with temperatures ranging up to 50 MeV or about $5 \cdot 10^{11}$ K. The core has to get rid of this heat in order to become a neutron star. The only way to do so is by emitting neutrinos. Photons have much too strong an interaction with ordinary matter: they can get out of the star only when the matter itself is ejected.

The core settles down into a sphere with a radius of some 20 km in which neutrinos diffuse in much the same way as photons do in ordinary stars. From the surface of this sphere neutrinos are emitted. It is called the neutrino sphere, in analogy to the photosphere of an ordinary star. At the neutrino sphere, the density is of the order of 10^{11} g cm^{-3}, and the temperature of the order of 5 MeV. Beyond this sphere, the neutrinos essentially move freely. However, they still have a certain chance to be absorbed by protons and neutrons, and give their energy to the matter in the process. Thereby this matter gets heated, develops pressure and drives the shock out.

This mechanism of Wilson gave successful supernova shocks. A characteristic feature was the formation of a bubble, a large region between neutrino sphere and shock front which contained very little matter, but had high pressure caused by a high intensity of electromagnetic radiation and electron pairs. This bubble propels the shock.

While the Wilson theory gave a successful shock, it had three drawbacks:

1. The shock was successful in some computer runs, but not in others. In 1989, Wilson and Mayle concluded that the flux of neutrinos from the core was not quite sufficient; so they artificially increased it by about 50% by postulating "salt finger" convection inside the core. As later developments have shown, this postulate is no longer necessary.
2. The explosion energy was found to be about 0.4 foe, whereas the observations on Supernova 1987A indicate 1.5 foe; so the mechanism gives insufficient energy.
3. The bubble has very low density, much lower than the material outside which is to be propelled in the shock; density ratios of 10 or more are common. Such a configuration is subject to Rayleigh–Taylor instability.

The third drawback has led to resolution of the problem. Rayleigh–Taylor instability itself does not work: it would lead to mixing of the material on the two sides of the interface between the bubble and the outside, but these materials are of the same kind — nucleons, electrons and photons. Instead, convection would take place. The condition for convection is that entropy decreases with increasing radius. This is well fulfilled: the entropy in the bubble equals several hundred while near the shock it is only of order 10.

As we know from observations on the sun as well as from laboratory experiments, convection takes place in cells; the hot material goes up on one side of the cell, the cold material down on the other. After moving a certain distance, the mixing length, the hot and the cold stream get mixed, and in the next cell above the same procedure occurs. In standard stellar evolution calculations, a very small entropy gradient suffices to drive convection. But in the supernova, where everything happens in seconds, much higher entropy gradients are needed. In an example, I found that the entropy (in units of k_B per nucleon) has to change by about 50 units from the shock to the neighbourhood of the neutrino sphere.

The first task for the neutrino heating plus convection is to turn the shock wave around. After the prompt shock stalls, material accretes to the core because the matter originally at large distance continues to fall in. This accreting material moves inward behind the shock, albeit with much smaller velocity than it had outside the shock. The rate of accretion determines the product

$$\frac{\lambda}{4\pi} = \rho r^2 v \,. \tag{4}$$

Knowing the mass distribution before the supernova collapse, and assuming more-or-less free fall, we find that

$$\rho r^2 v = 2 \cdot 10^{31} t^{-1} \text{ g s}^{-1} \,. \tag{5}$$

In computations, one finds typically that

$$v = 5 \cdot 10^8 \text{ cm s}^{-1} \,, \tag{6}$$

$$\rho r^2 = 4 \cdot 10^{22} \text{ g cm}^{-1} \,. \tag{7}$$

When the accreting material gets heated by neutrinos, the first task the heat has to accomplish is to turn the inward motion v into an outward motion. To examine the requirement for this, we use the Hugoniot relations,

$$(u_2 - u_1)^2 = (p_2 - p_1)(V_1 - V_2) \,, \tag{8}$$

$$2(E_2 - E_1) = (p_2 + p_1)(V_1 - V_2) \,. \tag{9}$$

Here quantities with subscript 1 refer to the outside of the shock, with subscript 2 to the inside. E is the internal energy per unit mass, p the pressure, $V = \frac{1}{\rho}$ the specific volume, and u the outward velocity, so $u_2 = -v$. We want to increase u_2 (algebraically), and since u_1 is large and negative, this means increasing $(u_2 - u_1)^2$. It is clear from (8) and (9) that this requires a sufficiently large $E_2 - E_1$. In order to make $u_2 > 0$, i.e. turn around the motion behind the shock, we must have

$$E_2 \geq E_{2\,\text{min}} = E_1 + \frac{1}{2} u_1^2 + p_1(V_1 - V_2) \,. \tag{10}$$

For a quantitative estimate, I used a computation by Wilson and Mayle in which the shock was at $R = 217$ km. It gave

$$E_{2\,\text{min}} = 10.6 \cdot 10^{18} \text{ ergs g}^{-1} \,, \tag{11}$$

while the actual internal energy was

$$E_2 = 9.2 \cdot 10^{18} \,. \tag{12}$$

Since $E_2 < E_{2\,\text{min}}$, $u_2 < 0$, i.e. the material behind the shock moves inward. The energies E_2 and $E_{2\,\text{min}}$ may be compared with the gravitational potential,

$$GM/R = 9.9 \cdot 10^{18} \,. \tag{13}$$

It is not surprising that E_2 is about equal to GM/R because the material outside the shock has acquired its internal and kinetic energy by falling down in the gravitational field. In order to turn the shock around, we need to raise the internal energy behind the shock by about 15%. This we propose to do by convecting energy from inside the shock.

According to the mixing length theory of convection, the velocity of the energy stream relative to the material is

$$v_c = \left(\frac{p}{\rho}\frac{d\ln S}{d\ln p}\right)^{\frac{1}{2}} . \tag{14}$$

The first factor in this is essentially the sound velocity. The entropy gets built up by neutrino absorption. The first requirement is that v_c becomes greater than v, the inward velocity of the material. Once this happens, the entropy generated by neutrino absorption moves outward rather than inward. I found that this requires raising S by 5 units, which can be accomplished in about 0.05 seconds.

The second requirement is to bring enough energy to the inside of the shock to turn it around. According to our discussion above, 0.2 GM/R per unit mass should be enough. Multiplied by the accretion rate λ, this is about 0.5 foe/s. It requires an entropy maximum about 10 units above the entropy behind the shock.

There is one process which interferes with the buildup of energy and entropy: the capture of electrons by nuclei. The rate of energy loss due to this process is

$$-\dot{w}_e = 2.0 \cdot 10^{18} T^6 \text{ ergs g}^{-1}\text{ s}^{-1} . \tag{15}$$

Here a factor T^4 arises from the energy density of electrons and positrons, and a factor T^2 from the cross section. The energy absorbed from neutrinos is

$$\dot{w}_\nu = 5.3 \cdot 10^{19} F_{52} r_7^{-2} \text{ ergs g}^{-1}\text{ s}^{-1} , \tag{16}$$

where F_{52} is the flux of neutrinos in units of 10^{52} ergs s^{-1}. Energy gain from neutrinos equals energy loss by electron capture when

$$T^3 r_7 = 5.1\ F_{52}^{\frac{1}{2}} . \tag{17}$$

We call this the gain point, R_g. At larger r the temperature T decreases rapidly, so the energy loss by electron capture is less than the energy gain. The material in the star may then be divided into three regions: (1) from the neutrino sphere to the gain point, (2) from the gain point to the entropy maximum, and (3) from that maximum to the shock. In region I all the energy gained from the neutrinos is lost by electron capture, in region II the energy gain dominates and entropy is built up, and in region III there is convection. All the energy generated in region III is convected to the shock front. So the *energy* available to the *supernova* shock is just the *total energy supplied by neutrinos to the region beyond the entropy maximum* (corrected for electron capture).

The energy gain, per unit neutrino flux, is

$$\alpha = F^{-1} \int \dot{w}_\nu d\tau = 5.3 \cdot 10^{-33} 4\pi \int \rho r^2 dr\, r_7^{-2} = 6.7 \cdot 10^{-18} \int \rho dr \,. \tag{18}$$

The radial integral over density can be calculated separately for small and for large shock radius R. For a small shock radius, $R = 300$ km, I assume that the density behaves as r^{-2}, as in Eq. (7); then

$$\int \rho dr \simeq \rho_g r_g \,. \tag{19}$$

This product can be calculated by combining (17) with the equation of state of a gas of electromagnetic radiation, which gives

$$\rho_g r_g = 2.6 \cdot 10^{16} \frac{F_{52}^{\frac{1}{2}}}{S_r} \,. \tag{20}$$

Using $F_{52} = 7$ and $S_r = 10.5$, and making a small correction for electron capture, we find that

$$\alpha = 2.4\% \,. \tag{21}$$

For a large shock radius, $R = 4000$ km, I have solved the equations of equilibrium explicitly. One of them is the hydrostatic equation

$$\frac{dp}{dr} = -\frac{GM}{r^2} \rho \,, \tag{22}$$

and the other gives the energy current due to convection,

$$J = 16\pi p r^2 \left(\frac{p}{\rho}\right)^{\frac{1}{2}} \left(\frac{d \ln S}{d \ln p}\right)^{\frac{3}{2}} . \tag{23}$$

I assumed that the pressure is dominated by radiation; then

$$\rho = 1.40 \cdot 10^{-11} \frac{p^{\frac{3}{4}}}{S} . \tag{24}$$

The result of the integration is given in Table 1. The table shows that ρr is nearly constant, so

$$\int \rho dr \simeq 0.35 \cdot 10^{15} \ln \left(\frac{R}{r_g} \right) . \tag{25}$$

Thus neutrino absorption occurs about equally, all over the region from the gain point to the shock. Putting in all corrections, we find that

$$\alpha = 0.7\% . \tag{26}$$

Averaging the two results, at small and large shock radius, we get

$$\alpha_{\text{av}} = 1.6\% . \tag{27}$$

Table 1. Distribution of interesting quantities for shock radius of 4000 km.

r(km)	$p_{24} r_8^2$	$\rho_8 r_7$	S	T (MeV)	$w_{24} r_8^2$
4000	20	0.67	10	0.32	45
2000	10.2	0.50	11.3	0.38	19.6
1000	6.4	0.40	13.9	0.48	10.2
500	4.6	0.35	18.1	0.62	6.1
250	4.0	0.32	24.7	0.84	5.0
125	3.6	0.32	33	1.17	3.8
62.5	3.7	0.35	43	1.65	3.4
31.2	4.0	0.42	53	2.4	2.8
15.6	5.6	0.63	64	3.7	3.0

The total gravitational energy liberated when a mass of 1.6 $M_\odot$ collapses into a neutron star is about $3 \cdot 10^{53}$ ergs. This energy is emitted about equally in each of the three neutrino flavours, ν_e, ν_μ and ν_τ; thus $1 \cdot 10^{53}$ erg is in $\nu_e + \bar{\nu}_e$. This is confirmed by the measurement of the $\bar{\nu}_e$ from Supernova 1987A in Japan and Ohio. Then the predicted supernova energy is

$$E_{\text{th}} = 1.6 \cdot 10^{51} \text{ ergs} . \tag{28}$$

The best number from the observation of the Doppler effect of the light from 1987A is

$$E_{\rm obs} = 1.5 \cdot 10^{51} \text{ ergs}\,. \tag{29}$$

The agreement is very good.

The computation in Table 1 also gives the energy per unit volume, internal plus gravitational:

$$w = 3p - \left(\frac{GM}{r}\right)\rho\,. \tag{30}$$

This is seen to be positive for all r, indicating that all the material involved can escape from the star. This is all the material which is involved in the convection. Thus, the separation of ejecta from the neutron star should take place at the entropy maximum, or slightly inside it.

The energy density w can now be integrated over the volume, and the result is

$$E = 4\pi \int wr^2 dr = 4\pi R^3 p_s\,, \tag{31}$$

where p_s is the pressure at the shock. This can actually be proved, for $\gamma = 4/3$, by the virial theorem. Since $p_s = w_i/3$, with w_i the energy density,

$$E = \frac{4\pi}{3} R^3 w_i\,. \tag{32}$$

Thus the internal energy behind the shock is equal to the total energy in the ejecta, divided by the volume inside the shock. This relation has been used for many years in studies of nucleosynthesis.

Nucleosynthesis takes place in material which, before the supernova event, was ^{16}O, ^{28}Si or ^{32}S. Therefore, synthesis leads to the most strongly bound nucleus with equal numbers of neutrons and protons. This is ^{56}Ni. As the supernova expands, this nucleus decays into ^{56}Co and this in turn into ^{56}Fe. Ni is relatively short-lived (6 days), but Co has a half-life of 77 days. Its radioactive decay powered the light emission from SN 1987A after the first month or two, and indeed for a long time the light emission followed accurately a 77-day decay time. The two principal γ-rays from ^{56}Co, 0.84 and 1.25 MeV, were observed.

As we have seen, the supernova is powered by the absorption of neutrinos. These are released from the proto-neutron star rather slowly, in several seconds, as we know both from theoretical studies of their diffusion and from the observations in the neutrino detectors at Kamiokande and IMB. Therefore the energy of the shock wave also builds up rather slowly, over several seconds. But its velocity is quite high from the beginning, more than 10^9 cm/s (10^4 km/s), because it runs into very dilute material.

I believe we now understand qualitatively how a supernova of type II works. The way is now open to investigate many important questions. For example, How does the mechanism work as a function of the mass of the original star? What are the details of nucleosynthesis (which, after all, provides all the elements beyond He, including all those we live on)?

References

1. E. A. Baron, J. Cooperstein and S. Kahana, *Nucl. Phys.* **A440**, 744 (1985).
2. E. A. Baron and J. Cooperstein, *Ap. J.* **353** (1990).
3. S. W. Bruenn, *Ap. J.* **340**, 955; **341**, 955 (1989).
4. J. R. Wilson and R. L. Bowers, *Ap. J.* **263**, 366 (1982).

THE BIG BANG AND COSMIC INFLATION

ALAN H. GUTH*
Center for Theoretical Physics
Laboratory for Nuclear Science and Department of Physics
Massachusetts Institute of Technology
Cambridge, Massachusetts 02139, USA

A summary is given of the key developments of cosmology in the 20th century, from the work of Albert Einstein to the emergence of the generally accepted hot big bang model. The successes of this model are reviewed, but emphasis is placed on the questions that the model leaves unanswered. The remainder of the paper describes the inflationary universe model, which provides plausible answers to a number of these questions. It also offers a possible explanation for the origin of essentially all the matter and energy in the observed universe.

1. Introduction

It is a pleasure to visit Stockholm and to speak in honor of Oskar Klein, certainly one of the great theoretical physicists of the 20th century. All physicists are familiar with the many important developments that are identified with Klein's name, such as the *Klein*–Gordon equation, the *Klein* paradox, the *Klein*–Nishina formula, the Kaluza–*Klein* theory, and *Klein*–Jordan second quantization.

However, although these contributions are extraordinarily impressive, I have to admit that Oskar Klein is not nearly as well known as he should be, especially among the younger generation of physicists. I believe that many of these physicists, at least in my own country, could not tell you what Klein's first name was. And many are not even aware that all of the items on the list above were in fact done by the same person.

*This work was supported in part by funds provided by the US Department of Energy (DOE) under contract #DE-AC02-76ER03069.

One cannot be certain how to explain this failure in public relations, but I have a theory, and a proposal to solve the problem. My theory is that the name "Klein" is just too simple, too short, too generic and too prosaic to capture anyone's attention. To investigate this theory scientifically, I did some sociological research — in the Boston telephone directory.[1] (Boston is clearly one of the most unusually average cities of the world, and in any case I had a copy of the Boston telephone directory in my closet.) The directory lists no less than 179 Klein's, ranging alphabetically from Adam Klein to Yasue Klein. Testing against a carefully selected control group, my research showed that there are 0 Gell-Mann's, 0 Feynman's, 0 Dirac's, 0 Einstein's, 0 Bohr's, 0 De Broglie's, 0 Schrödinger's, and only 1 Guth (i.e. me). Clearly, Prof. Klein's hard-working public relations agents have a serious problem to overcome.

In the face of this relentless uphill battle, I suggest that the situation can be significantly ameliorated by asking a simple question: if Oskar Klein were alive today, what would he use for a bitnet address? I contend that he would very likely do what so many of my friends with common last names have done — that is, he would append his first initial to his last name, giving

Oklein@sesuf51

If we all agree to follow this usage consistently, we can add the following *distinctive* terms to the physicists' vocabulary: the *Oklein*–Gordon equation, the *Oklein* paradox, the *Oklein*–Nishina formula, Kaluza–*Oklein* theory, and *Oklein*–Jordan second quantization. I have checked out this proposal with my professional friends on Madison Avenue, and they confirm that the strategy is likely to improve Prof. Klein's name recognition by at least 86%.

So, having settled that issue, I would now like to move on to the main topic of this talk, a subject which was of great interest to Oskar Klein: the very early history of the universe. First, I should explain that while I will be talking about cosmology, the perspective will be that of a particle theorist. In the late 1970's, I joined a small drove of particle theorists who began to dabble in studies of the early universe. We were motivated partly by the intrinsic fascination of cosmology, but also by new developments in elementary particle physics itself.

The particle physics motivation arose primarily from the advent of a new class of theories, known as "grand unified theories." These theories were invented in 1974, but it was not until about 1978 that they became a topic of widespread interest in the particle physics community. Spectacularly bold, these theories attempt to extend our understanding of particle physics to energies of about 10^{14} GeV (1 GeV = 1 billion electron volts $\approx$ rest energy of a proton). This amount of energy, by the standards of your local power company, may not seem so impressive — it is about what it takes to light a 100-watt bulb for one minute. The grand unified theories, however, attempt to describe what happens when that much energy is deposited on a single elementary particle. Such an extraordinary concentration of energy exceeds the capabilities of the largest existing particle accelerators by 11 orders of magnitude.

To get some feeling for how high this energy really is, imagine trying to build an accelerator that might reach these energies. One can do it in principle by building a very long linear accelerator. The largest existing linear accelerator is at Stanford, with a length of about 2 miles and a maximum energy of about 50 GeV. The output energy is proportional to the length, so a simple calculation shows how long an accelerator would have to be to reach an energy of 10^{14} GeV. The answer is almost exactly one light-year!

Unfortunately, both the US Department of Energy and NASA seem to be very unreceptive to proposals for funding a one-light-year-long accelerator. Consequently, if we want to see the most dramatic new implications of the grand unified theories, we are forced to turn to the only laboratory, to which we have any access, that has ever reached these energies. That "laboratory" appears to be the universe itself, in its very infancy. According to the standard hot big bang theory of cosmology, the universe had a temperature corresponding to a mean thermal energy of 10^{14} GeV at a time of about 10^{-35} second after the big bang. No wonder particle theorists suddenly became interested in the very early universe.

The first half of this lecture will review the standard hot big bang model of the early universe, while the second half will discuss developments that have taken place since 1978 — developments which have been motivated mainly by ideas from particle physics.[2]

2. The Big Bang Theory

Cosmology in the 20th century began with the work of Albert Einstein.[3] In March 1916, Einstein completed a landmark paper titled "The Foundation of the General Theory of Relativity."[4] The theory of general relativity is, in fact, nothing more or less than a new theory of gravity. Complex but very elegant, the theory describes gravity as a distortion of the geometry of space and time. Unlike Newton's theory of gravity, general relativity is consistent with the ideas of special relativity, which Einstein had introduced in 1905. While the rest of the world waited to be persuaded, Einstein was immediately convinced that he had found the correct description of gravity.

In less than a year after the publication of general relativity, Einstein applied it to the universe as a whole.[6] However, in carrying out these studies Einstein discovered something that surprised him a great deal: it was impossible to build a static model of the universe consistent with general relativity. Einstein was perplexed by this fact. Like his predecessors, he had looked into the sky, saw that the stars appeared motionless, and erroneously concluded that the universe is static.

In fact, the same problem that Einstein discovered in the context of general relativity also existed in Newtonian mechanics, although it had not been fully understood until the work of Einstein. The problem is nonetheless fairly simple to understand: if masses were distributed uniformly and statically throughout space, then everything would attract everything else and the entire configuration would collapse. (Netwon himself had wrestled with this problem, but had incorrectly convinced himself that, if the universe is infinite, the forces in opposite directions would cancel and the collapse would be avoided.[8]) Einstein nonetheless remained convinced that the universe was static. He therefore modified his equations of general relativity, adding what he called a "cosmological term" — a kind of universal repulsion that prevents the uniform distribution of matter from collapsing under the normal force of gravity. Of course, if one is constructing a theory of gravity, one does not want to introduce a repulsive force that cancels the attractive force under all circumstances. The cosmological force, however, has a different distance dependence than the attractive force. In the Newtonian approximation, the attractive force is proportional to $1/r^2$, where r is the distance, while the cosmological force grows with distance, proportional to r. In this way the cosmological force can stabilize the universe against collapse, while its effects are negligible on the scale of the solar system or smaller. The cosmological term, Einstein found, fits neatly into the equations of general relativity — it is completely consistent with all the fundamental ideas on which the theory was constructed.

Einstein's picture of cosmology remained viable for about a decade, until astronomers began to measure the velocities of distant galaxies. They then discovered that the universe is not at all static. To the contrary, the distant galaxies are receding from us at high velocities.

The pattern of the cosmic motion was codified at the end of the 1920's by Edwin Hubble, in what we now know as Hubble's law.[9] Hubble's law states that each distant galaxy is receding from us with a velocity which is, to a high degree of accuracy, proportional to its distance. Thus one can write

$$v = Hl \, ,$$

where v is the recession velocity, l is the distance to the galaxy, and the quantity H is known as the Hubble "constant." (I put the word "constant" in quotation marks to call attention to its inaccuracy. The quantity was called a "constant" by astronomers, presumably because it remains approximately constant over the lifetime of an astronomer. The value of H changes, however, as the universe evolves; so, from the point of view of a cosmologist, it is not a constant at all.) Hubble's law is not a precise law of nature, but the observed velocities of galaxies are generally found to be very close to the values predicted by Hubble's equation.

The value of the Hubble constant is not very well known. The recession velocities of the distant galaxies are no problem—they can be determined very accurately from the Doppler shift of the spectral lines in the light coming from the galaxies. The distances to the galaxies, on the other hand, are very difficult to determine. These distances are estimated by a variety of indirect methods, and the resulting value of the Hubble constant is thought to be uncertain by a factor of about two. It is believed to lie somewhere in the range

$$H \approx \frac{0.5 \text{ to } 1}{10^{10} \text{ years}} .$$

Notice that the Hubble constant has the units of inverse time; when the Hubble constant is multiplied by a distance, the result has the units of distance per time, or velocity. In particular, if the expression for H above is multiplied by a distance in light-years, the result is a velocity measured as a fraction of the velocity of light. Alternatively, one can use

$$H \approx (15 \text{ to } 30) \text{ km/sec per million light-years}$$

to obtain a velocity in kilometers per second.[a]

The development of cosmology in the 20th century was somewhat confused by the fact that Hubble badly overestimated the value of the Hubble constant, reporting a value of 150 km/sec per million light-years. This mismeasurement had important consequences. In the context of the big bang model, an erroneously high value for the expansion rate implies an erroneously low value for the age of the universe. Hubble's value for the Hubble constant implied an age of about two billion years, a number which conflicts with geological evidence that the Earth is significantly older. This contradiction no doubt

[a]Astronomers generally quote the Hubble constant in km/sec per Megaparsec, where 1 Megaparsec = 3.26 million light-years. In these units the Hubble constant lies between 50 and 100.

contributed to the popularity of the steady-state model during the middle of the 20th century. Not until 1958 did the measured value of the Hubble constant come within the currently accepted range, due primarily to the work of Walter Baade[10] and Allan Sandage.[11] The age of the universe is now estimated to be between 10 and 20 billion years.

Once it is noticed that the other galaxies are receding from us, there are two conceivable explanations. The first is that we might be in the center of the universe, with everything moving radially outward from us like the spokes on a wheel. In the early 16th century such an explanation would have been considered perfectly acceptable. Since the time of Copernicus, however, astronomers and physicists have become instinctively skeptical of this kind of inference, so alternative explanations are sought. In this case, an attractive alternative can be found.

The alternative explanation can be called *homogeneous expansion*, and it is illustrated in Fig. 1. The three pictures are intended to show successive snapshots of a region of the universe. Each picture is essentially a photographic blowup of the previous picture, with all distances enlarged by the same percentage. According to this explanation, there is nothing special about our galaxy — or any other galaxy. All galaxies are approximately equivalent, and are spread more or less uniformly throughout all of space. (In this description, there is no center and no edge to the distribution of galaxies.) As the system evolves, from diagram (a) to (b) to (c), *all* intergalactic distances are enlarged. Thus, regardless of which galaxy we are living on, we would see all the other galaxies receding from us. Furthermore, this picture leads immediately to the conclusion that the recession velocities obey Hubble's law. Since all distances increase by the same *percentage* as the system evolves, larger distances increase by a larger amount. The apparent velocity of a galaxy is proportional to the amount by which the distance from us increases, and hence it is proportional to the distance.

The homogeneous expansion can be described mathematically in a simple way by introducing what is called a "comoving" coordinate system. This is a coordinate system which is marked off with grid marks that expand with the universe, as shown in Fig. 1. A typical galaxy will then maintain constant values of its x, y, and z coordinates as the universe expands. The comoving coordinate system thus provides a map of the universe that remains accurate even as the universe expands. If the universe doubles in size, one simply finds the words "1 grid mark = 25 light-years" in the lower right corner and replaces them with "1 grid mark = 50 light-years." The expansion is thereby

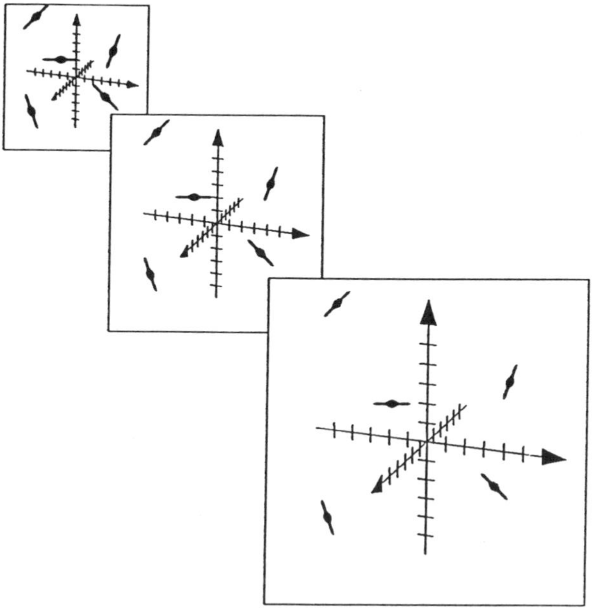

Fig. 1. The expanding homogeneous universe. The three pictures show snapshots of the same region of the universe, taken at three successive times. Each picture is essentially a photographic blowup of the previous picture, with all the distances enlarged by the same percentage. An observer on any galaxy would conclude that all the other galaxies are receding from him.

incorporated into an overall scale factor, R, which depends on time:

$$l_{\text{phys}}(t) = R(t) l_{\text{coord}} ,$$

where l_{coord} is the (time-independent) distance between any two galaxies measured in grid marks, $R(t)$ is the scale factor, and $l_{\text{phys}}(t)$ is the physical distance between the two galaxies. When the value of $R(t)$ doubles, it means that all of the distances in the universe have doubled. (Since the Hubble expansion pattern is not rigorously exact, the comoving coordinates of real galaxies are not exactly time-independent, but instead vary slowly with time.)

Figure 1 shows that while the universe is expanding, the size of the individual galaxies does not increase with time. It is believed that shortly after the big bang, the matter that makes up the galaxies was in the form of a nearly uniform gas undergoing nearly uniform expansion. The galaxies then formed from the gravitational clumping of matter, and their size is now stabilized by the dynamics of internal motions and gravitational binding.

When one extrapolates this picture backward in time, one finds an instant in the past when all the galaxies must have been on top of each other and when the density of the universe would have been infinite. Such an event is called a "singularity," and in this case the singularity is the instant of the "big bang" itself, some 10 to 20 billion years ago. The time of the big bang is very uncertain for two reasons: we do not know the Hubble constant very accurately, and we are also uncertain about the mass density of the universe. (The mass density is important in calculating the history of the universe, because it determines how fast the cosmic expansion is slowing down under the influence of gravity.)

The reader should be warned, however, that the calculation implying an infinite density at the instant of the big bang is not to be trusted. As one looks backward in time with the density going up and up, one is led further and further from the conditions under which the laws of physics as we know them were developed. Thus, it is quite likely that at some point these laws become totally invalid, and then it is a matter of guesswork to discuss what happened at earlier times. Nonetheless, as the history of the universe is extrapolated backward, the density increases without limit for as long as the known laws of physics apply. Indeed, most cosmologists today are reasonably confident in our understanding of the history of the universe back to one microsecond ($t \approx 10^{-6}$ second). The goal of the cosmological research involving grand unified theories is to extend our understanding back to $t \approx 10^{-35}$ second.

The time scale that I am using is defined so that the extrapolated infinite density occurs at $t \equiv 0$. Even though the extrapolation to infinite density is not reliable, it still seems to be the most convenient way to define what we mean by $t = 0$. Despite this uncertainty about the zero point of time, we still believe that the time scale used in cosmology gives a meaningful measure of the time intervals between those events that are described. (The inflationary theory, as I will discuss later, gives a much more complicated picture of what happened at $t = 0$, *and even earlier.*)

Having discussed the key features of the big bang theory, we can now ask what evidence can be found to support it. In addition to Hubble's law, there are two significant pieces of observational evidence in favor of the big bang theory.

The first is the observation of the cosmic background radiation. To understand the origin of this radiation, begin by recalling that the temperature of a gas rises when the gas is compressed. For example, a bicycle tire is warmed when it is inflated by a hand pump. Similarly, a gas cools when it is allowed to expand. In the big bang model, the universe has been expanding throughout its history, which means the early universe must have been much hotter. (In fact, the standard calculation implies that the temperature was infinite at the instant of the big bang. This calculation, however, like the calculation of the infinite density, should not be taken as definitive.)

All hot matter emits a glow, just like the glow of hot coals in a fire. Thus, the early universe would have been permeated by the glow of light emitted by the hot matter. As the universe expanded, this light would have cooled, and hence red-shifted (i.e. decreased in frequency). Today, the universe would still be bathed by the radiation, a remnant of the intense heat of the big bang—now red-shifted into the microwave part of the spectrum. This prediction was confirmed in 1964 when Arno A. Penzias and Robert W. Wilson[12] of the Bell Telephone Laboratories discovered a background of microwave radiation with an effective temperature of about 3°K.

At the time of their discovery, Penzias and Wilson were not looking for cosmic background radiation. Instead, they were searching for astronomical sources capable of producing low-level radio interference. They discovered a hiss in their detector, and found that the hiss was present no matter what direction in the sky the antenna was pointed. The hiss seemed likely to be merely electrical noise from the detector, but Penzias and Wilson carefully examined all sources of noise, concluding that the signal was really coming from the sky.

Penzias and Wilson measured the intensity of the radiation at only one frequency, but other experimenters set out to determine whether the spectrum of the radiation agrees with the thermal spectrum predicted by the big bang theory. The data on the spectrum available in the early 1980's is summarized in Fig. 2, a graph adapted from *The Big Bang*, by Joseph Silk.[13] The diagram shows the original point found by Penzias and Wilson, and also a number of other more recent measurements. The solid line is the spectral curve expected for thermal radiation, in this case at a temperature of 2.9°K. It is clear that the data agrees well with the theoretical prediction, but note that there are at this point no measurements on the steep part of the curve on the high frequency side (at the right).

The earth's atmosphere poses a serious problem for measuring the high frequency side of the curve, so the best measurements must be done from balloons, rockets, or satellites. In 1988 a rocket probe was launched by a collaboration between the University of California at Berkeley, and Nagoya University in Japan.[14] Their results are shown in Fig. 3. Note that the points labeled 2 and 3 are much higher than the thermal spectrum predicts. Fitting

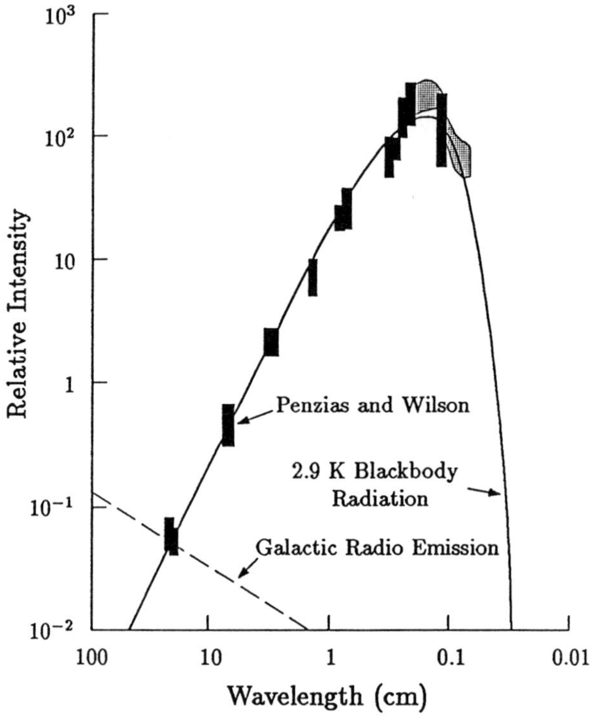

Fig. 2. The cosmic background radiation. The vertical lines and the shaded region show the results of intensity measurements at a variety of wavelengths. The data shown here was all gathered before 1981. The uncertainties are large, but the data is consistent with the theoretical curve, corresponding to thermal radiation at a temperature of 2.9°K. Note that there are no measurements on the right of the curve, where the theoretical prediction shows a very sharp drop in intensity.

these points individually to a temperature, the authors find:

$$\text{Point 2:} \quad T = 2.955 \pm 0.017°\text{K},$$
$$\text{Point 3:} \quad T = 3.715 \pm 0.027°\text{K}.$$

These numbers correspond to discrepancies of 12 and 16 times the estimated uncertainty, respectively, from the temperature of $T = 2.74°$K which fits the lower frequency points. In terms of energy, the excess intensity seen at high frequencies in this experiment amounts to about 10–15% of the total energy in the cosmic background radiation. Cosmologists were stunned by the extremely significant disagreement with predictions. Some tried to develop theories to

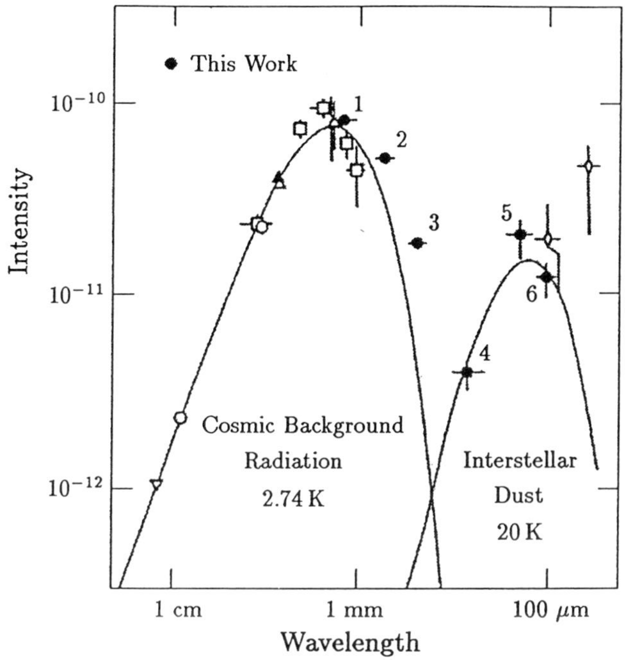

Fig. 3. The submillimeter excess background radiation. The Berkeley–Nagoya rocket experiment of 1988 measured the cosmic background radiation at the high frequency end of the spectrum, and found significantly more intensity than was expected. The discrepancy is seen in the points labeled 2 and 3, which lie far above the theoretically expected curve. (The vertical axis shows frequency times intensity per unit frequency, in units of watts per cm^2 per steradian.)

explain the radiation, without much success, while others banked on the theory that it would go away. The experiment looked like a very careful one, however, so it was difficult to dismiss. The most likely source of error in an experiment of this type is the possibility that the detectors were influenced by heat from the exhaust of the launch vehicle — but the experimenters very carefully tracked how the observed radiation varied with time as the detector moved away from the launch rocket, and it seemed clear that the rocket was not a factor.

The same group tried to check their results with a second flight a year later, but the rocket failed and no useful data was obtained.

In the fall of 1989 NASA launched the Cosmic Background Explorer, known as COBE. This marked the first time that a satellite was used to probe the background radiation. Within months, the COBE group announced their first results at a meeting of the American Astronomical Society in Washington, DC, January 1990. The detailed preprint, with a cover sheet showing a sketch of the satellite, was released the same day.

The data showed a perfect fit to the thermal spectrum,[15] with a temperature of $2.735 \pm 0.06°$ K, and no evidence whatever for the "submillimeter excess" that had been seen by Matsumoto *et al.* The data was shown with estimated error bars of 1% of the peak intensity, which the group regarded as very conservative. The graph is shown here as Fig. 4.

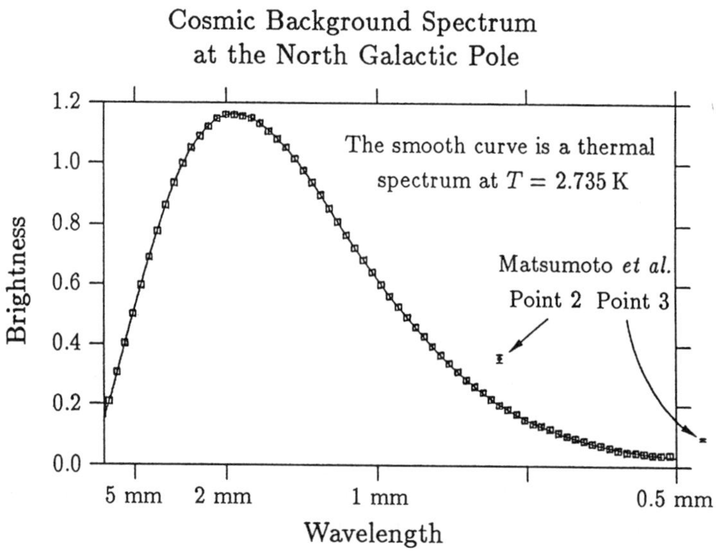

Fig. 4. The cosmic background radiation as seen by COBE. By far the most precise determination of the spectrum to date indicates that it agrees beautifully with the thermal spectrum predicted by the big bang. The earlier points measured by Matsumoto *et al.* are shown on the same scale for comparison. (The vertical axis shows the intensity per unit frequency times the speed of light, in units of 10^{-11} watts per cm^2 per steradian per hertz, times cm/sec.)

The COBE graph is plotted on a completely different scale from the graph presented by Matsumoto *et al.*, so they are difficult to compare. I have therefore added the points 2 and 3 of the Matsumoto *et al.* data to the COBE graph. Since the COBE instrument is far more precise and has more internal consistency checks, there has been no doubt in the scientific community that the COBE result supersedes the previous one. Despite the 16σ discrepancy of 1988, the cosmic background radiation is now once again believed to have a nearly perfect thermal spectrum.

The second important piece of evidence supporting the big bang theory is related to calculations of what is called "big bang nucleosynthesis." This evidence is slightly more difficult to understand than the cosmic background radiation, and it is therefore much less discussed in popular scientific literature. To make sense of this argument, one must understand that the big bang theory is not just a cartoon description of a universe beginning with a big explosion. On the contrary, the big bang theory is a very detailed model. Once one accepts its basic assumptions, then knowledge of the laws of physics allows one to calculate how fast the universe would have expanded, how fast the expansion would have been slowed by gravity, how fast the universe would have cooled, and so on. Given this information, knowledge from nuclear physics allows one to calculate the rates of the different nuclear reactions that took place in the early history of the universe.

The early universe was very hot, so hot that even the nuclei of atoms would not have been stable. At two minutes after the big bang there were virtually no nuclei at all. The universe was filled with a hot gas of photons and neutrinos, with a much smaller density of protons, neutrons, and electrons. (The protons, neutrons, and electrons were very unimportant at the time, but they later became raw materials for the formation of stars and planets.) As the universe cooled, the protons and neutrons began to coalesce to form nuclei. From the nuclear reaction rates, one can calculate the expected abundances of the different types of nuclei that would have formed. One finds that most of the matter in the universe would remain in the form of hydrogen. About 25% (by mass) of the matter would have been converted to helium, and trace amounts of other nuclei would also have been produced.

Most of the types of nuclei that we observe in the universe today were produced much later in the history of the universe, in the interiors of stars and in supernova explosions. When a star undergoes a supernova explosion, its material is strewn into space, ready to recollect into further generations of stars. The lightest nuclei, however, were produced primarily in the big bang, and it is possible to compare the calculated abundances with direct observations. Such a comparison can be carried out for the abundances of helium-4, helium-3, hydrogen-2 (otherwise known as deuterium), and lithium-7.

The comparison is complicated by the fact that we do not know all the information necessary to carry out the calculations. In particular, the calculation depends sensitively on the density of protons and neutrons in the universe, a quantity which can be estimated only roughly by astronomical observations. Thus, the calculations have to be carried out for a broad range of values for

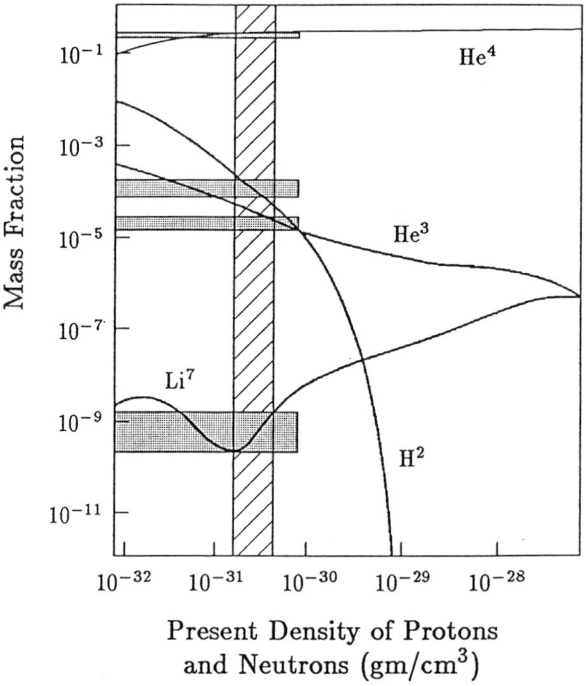

Fig. 5. Big bang nucleosynthesis. The curves show how the predicted abundances of the light chemical elements — helium-4, helium-3, deuterium (hydrogen-2), and lithium-7 — depend on the present density of protons and neutrons. The shaded horizontal bars show the observed abundances, with the estimated uncertainties. Excellent agreement is obtained for the range of densities indicated by the diagonally striped region.

this density. One then asks whether there exists a plausible value for which the answers turn out right.

The results of this comparison[16,17] are shown in Fig. 5. The horizontal axis shows the present density of protons and neutrons, and the curves indicate the results of the calculation. The observations, with the estimated range of their uncertainties, are shown by the shaded horizontal bars. Note that there is a range of values for the density of protons and neutrons, indicated by the diagonally striped region, for which each of the four calculated curves agrees with the corresponding observation. Even though the abundances of these nuclei are not known to high precision, the success of the comparison is very impressive. Note that the abundances span nine orders of magnitude. If there was never a big bang, there would be no reason whatsoever to expect that helium-4 would be 10^8 times as abundant as lithium-7 — it might just as well have been the other way around. But, when calculated in the context of the big bang theory, the ratio works out just right.

3. Unanswered Questions

Each piece of evidence discussed above — Hubble's law, the cosmic background radiation, and the big bang nucleosynthesis calculations — probes the history of the universe at a different period of time. The observation of Hubble's law, for example, probes the behavior of the universe at times comparable to the present — billions of years after the big bang. The cosmic background radiation, on the other hand, samples the conditions in the universe about 300,000 years after the big bang, when the universe became cool enough for the plasma of free nuclei and electrons to condense into neutral atoms. The plasma that filled the universe at earlier times was almost completely opaque to photons, which would have been constantly absorbed and re-emitted. With the formation of neutral atoms, however, the universe became highly transparent. Thus, most of the photons in the cosmic background radiation have been moving in a straight line since 300,000 years after the big bang, and they therefore provide an image of the universe at that time. Finally, the big bang nucleosynthesis calculations probe the history of the universe at much earlier times. The processes involved in establishing the abundances of the light nuclei occurred at times ranging from about one second to about four minutes after the big bang.

The big bang theory is a very successful description of the evolution of the universe for the whole range of times discussed above — from about one second after the big bang to the present. Nonetheless, the standard big bang theory has serious shortcomings, in that a number of very obvious questions are left unanswered. Here I describe three of these questions, which are listed in Table 1. Later I will show how ideas from particle physics have led to a radically new picture for the very early behavior of the universe, a picture which provides plausible answers to each of these questions.

The first question is related to the large-scale homogeneity, or uniformity, of the observed universe. The discussion of homogeneity must be qualified, however, because the universe that we observe is in many ways very inhomogeneous. The stars, galaxies, and clusters of galaxies make a very lumpy distribution. Cosmologically speaking, however, all of this structure in the universe is very small-scale. If one averages over very large scales, scales of 300 million light-years or more, then the universe appears to be very homogeneous. This large-scale homogeneity is most evident in the cosmic background

Table 1. Questions left unanswered by the standard big bang theory.

#1:	How did the universe become so homogeneous on large scales? Do we have to assume that it started out that way?
#2:	Why was the mass density of the early universe so extraordinarily close to the critical density?
#3:	Can one find a physical origin for the primordial density perturbations which led to the evolution of galaxies and clusters of galaxies? Are there physical processes that determine the spectrum of these perturbations?

radiation. Physicists have probed the temperature of the cosmic background radiation in different directions, and have found it to be extremely uniform. It is just slightly hotter in one direction than in the opposite direction, by about one part in 1000. Even this small discrepancy, however, can be accounted for by assuming that the solar system is moving through the cosmic background radiation at a speed of about 600 km/sec. Once the effect of this motion is subtracted out, the resulting temperature pattern is uniform in all directions to extraordinary accuracy. Until 1992 no nonuniformities had been detected at all, with probes sensitive to a level of a few parts per 100,000. In April 1992 the COBE team announced the results of their nonuniformity measurements, based on the first year of data collected by the satellite. The first nonuniformities had been discovered,[18] at the extremely small level of only about 6 parts per million. Since the cosmic background radiation gives an image of the universe at 300,000 years after the big bang, one concludes that the universe was extremely homogeneous at that time. (The observed "small-scale" inhomogeneities are believed to have formed later, by the process of gravitational clumping.) The standard big bang theory cannot explain the large-scale uniformity; instead, the uniformity must be postulated as part of the initial conditions.

The difficulty in explaining the large-scale uniformity is a quantitative question, related to the rate of expansion of the universe. Under many circumstances a uniform temperature would be easy to understand — anything will come to a uniform temperature if it is left undisturbed for a long enough period of time. In the standard big bang theory, however, the universe evolves so quickly that it is impossible for the uniformity to be created by any physical process. In fact, the impossibility of establishing a uniform temperature depends on none of the details of thermal transport physics, but is instead a direct consequence of the principle that no information can propagate faster than the speed of light. The situation is illustrated in Fig. 6, which shows a

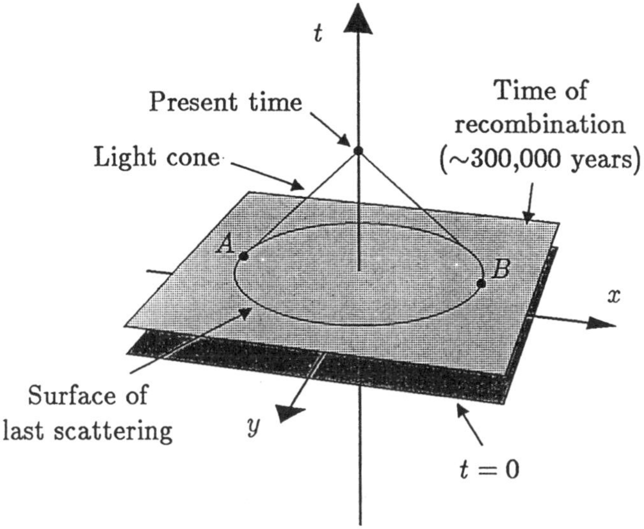

Fig. 6. A spacetime diagram for the universe. The time axis has been distorted so that light pulses travel at 45°. The early plasma-filled universe is opaque to radiation, but photons can travel freely starting at the surface labeled the time of recombination. Points A and B indicate the origins of cosmic background radiation photons that are arriving at earth today from two opposite directions in the sky. These points were separated from each other by about 100 times the distance that light could have traveled prior to that time.

spacetime diagram of the universe. The diagram illustrates two of the three dimensions of space, and also time. (The x and y axes of Fig. 6 represent comoving coordinates, as defined in the discussion of Fig. 1.) The time axis has been distorted and rescaled so that a light pulse travels in the diagram at 45° from the vertical. Our location in space and time is shown as a point on the time axis. The initial singularity is shown as a darkly shaded surface at $t = 0$. The early universe was opaque to radiation during its plasma phase, which ended at $t \approx 300{,}000$ years, shown as a lightly shaded surface labeled "time of recombination." ("Recombination" refers to the combining of nuclei and electrons to make neutral atoms; the meaning of the prefix "re" remains a mystery to the author.) The bulk of the photons in the cosmic background radiation arriving at earth today were last scattered at this time, and hence they originate on the circle labeled "surface of last scattering." (If the full three dimensions of space could be shown, this circle would become a spherical surface.) The spacetime points A and B represent the origins of two cosmic background radiation photons arriving at earth today from two opposite directions in the sky. The problem is that these two points are separated from each other by more than 90 times the distance that light could have traveled up until that time. (Note that for the sake of visibility the diagram has not been drawn to scale; if it were, the two shaded surfaces would be closer to each other by a factor of ten.) Thus, there is no way that points A and B could have communicated with each other, and certainly no physical process that would bring them to the same temperature.

The puzzle of explaining why the universe appears to be uniform over such large distances is not a genuine inconsistency of the standard theory: if the uniformity is assumed in the initial conditions, then the universe will evolve uniformly. The problem is that one of the most salient features of the observed universe — its large-scale uniformity — cannot be explained by the standard big bang theory; it must be assumed as an initial condition.

The second question is related to the mass density of the universe. This mass density is usually measured relative to a benchmark called the "critical mass density," which is defined in terms of the expansion rate of the universe.[b] If the mass density exceeds the critical density, then the gravitational pull of everything on everything else will be strong enough to eventually halt the expansion. The universe would recollapse, resulting in what is sometimes called a "big crunch." On the other hand, if the mass density is less than the critical density, the universe will go on expanding forever.

Cosmologists typically describe the mass density of the universe by a ratio designated by the Greek letter Ω (omega), defined by

$$\Omega \equiv \frac{\text{Mass density}}{\text{Critical mass density}} .$$

Ω is very difficult to determine, but its present value is known to lie somewhere in the range of 0.1 to 2.

That seems like a broad range, but consideration of the time development of the universe leads to a different point of view. $\Omega = 1$ is an *unstable equilibrium point* of the evolution of the standard big bang theory, which means it resembles the situation of a pencil balancing on its sharpened end. The phrase "equilibrium point" implies that if Ω is ever exactly equal to one, it will remain exactly equal to one forever[c] — just as a pencil balanced precisely on end

[b]The critical density is given by $3H^2/8\pi G$, where H is the Hubble constant and G is Newton's gravitational constant. For the range of values for H given earlier, the critical density lies in the range $0.4\text{–}1.8 \times 10^{-29}$ gm/cm^3.

[c]Since $\Omega = 1$ is a statement about the ultimate fate of the universe, it is not surprising that the condition is unchanged as time passes.

will, according to the laws of classical physics, remain forever in the vertical position. The word "unstable" means that any deviation from the equilibrium point, in either direction, will rapidly grow. If the value of Ω in the early universe is just a little bit above one, it will rapidly rise toward infinity; if Ω in the early universe is just a tiny bit below one, it will rapidly fall toward zero. It therefore seems very unlikely that the value of Ω today would lie *anywhere* in the vicinity of one.

Thus, if Ω is to be anywhere near one today, it must have been extraordinarily close to one at early times. For example, we can consider the time of one second after the big bang, the time at which the processes related to big bang nucleosynthesis were beginning to take place. In order for Ω to be somewhere in the allowed range today, at one second after the big bang Ω had to have been equal to one to an accuracy of 15 decimal places. In other words, at 1 second after the big bang, Ω was between 0.999999999999999 and 1.000000000000001. If we go further and consider the time of 10^{-35} second after the big bang, when thermal energies were typical of the energy scale of grand unified theories, then at that time Ω had to have been equal to one to an accuracy of 49 decimal places!

In the standard big bang theory there is no explanation whatever for this fact, as has been emphasized by Robert H. Dicke and P. James E. Peebles[19] of Princeton University. At one second after the big bang, Ω could have had any value — except that most possibilities would lead to a universe very different from the one in which we live. Like the large-scale homogeneity, the nearness of the mass density to the critical density cannot be explained; instead, it must be postulated as part of the initial conditions.

The third question concerns the origin of the density perturbations that are responsible for the development of the small-scale inhomogeneities. While the universe is remarkably homogeneous on the very large scale, there is nonetheless a very complicated structure on the scale of superclusters of galaxies and smaller. The existence of this structure is undoubtedly related to the instability of gravitational clumping: any region that contains a higher-than-average mass density will produce a stronger-than-average gravitational field, thereby pulling in even more excess mass. Thus, small perturbations are amplified to become large perturbations.

However, in order for galaxies to evolve, the early universe must have contained primordial density perturbations. The standard big bang model offers no explanation for either the origin or the form of these perturbations. Instead, an entire spectrum of primordial perturbations must be assumed as part of the initial conditions.

4. The Mechanism of Inflation

All three of the questions in Table 1 can be given plausible answers in terms of a model known as the inflationary universe. Before discussing these answers, however, I will have to spend some time describing how the model works. The description will be somewhat sketchy, but I will try to explain the main features.[20]

The inflationary universe was first proposed by me[29] in 1981, but the model in its original form did not quite work. It had a crucially important technical flaw, which was pointed out but not remedied in the original paper. A variation that avoids this flaw was invented independently by Andrei D. Linde[30] of the P. N. Lebedev Physical Institute in Moscow and by Andreas Albrecht and Paul J. Steinhardt[31] of the University of Pennsylvania. This section will discuss mainly the Linde–Albrecht–Steinhardt version of the model, which is called the *new* inflationary universe.

The mechanism of inflation depends on scalar fields, so I will begin by briefly summarizing the role of scalar fields in particle physics. To begin with, the reader should recognize that in the context of modern particle physics, *all* fundamental particles are described by fields. The best-known example is the photon. The classical equations describing the electromagnetic field were written down in the 1860's, but then in the early 20th century physicists learned that the underlying laws of nature are quantum and not classical. The quantization of the electromagnetic field can be carried out in a very straightforward way. For simplicity one can consider the fields inside a box, letting the size of the box approach infinity at the end of the calculation. One can then think of the electromagnetic fields inside the box as a mechanical system. The electromagnetic field is written as a sum of normal modes, and the coefficients of the normal mode functions can be taken as the dynamical degrees of freedom of the system. It turns out that each of these coefficients obeys the equations of a harmonic oscillator, and there is no interaction between the coefficients. The system is quantized by the same rules that one uses to quantize the hydrogen atom or the harmonic oscillator. The result is that each normal mode, like a harmonic oscillator, has evenly spaced energy levels. In this case, we interpret each energy level above the ground state as the occupation of the mode by a photon. Thus, the photon is interpreted as the quantized excitation of a field. This formalism is known as second quantization, and was of course pioneered by Oklein and Jordan.[32]

In contemporary particle theory, all elementary particles are described in this way. There is an electron field to correspond to an electron, a quark field to correspond to a quark, a neutrino field to correspond to a neutrino, etc. Among the different types of fields, the simplest is the scalar field — a field that has the same value to any Lorentz observer, and which evolves according to the Oklein–Gordon equation.[33] The quantized excitation of a scalar field is a spinless particle. Although spinless particles that are regarded as elementary have yet to be observed, they are nonetheless a key ingredient to a number of important theories. In particular, the Glashow–Weinberg–Salam model of the electroweak interactions makes use of a scalar field, called the Higgs field, to cause a symmetry in the theory to be spontaneously broken. (If this symmetry were not broken, then electrons and neutrinos would both be massless, and would be indistinguishable.) This Higgs field corresponds to a neutral spinless Higgs particle, which will hopefully be observed at the SSC (Superconducting Super-Collider), if not before. Grand unified theories make use of similar Higgs fields, but at a much higher mass scale, to spontaneously break the grand unified symmetry which relates electrons, neutrinos, and quarks.

There is much current interest in superstring theories, which actually go beyond the pattern of the field theories that were described above. In these theories the fundamental object is not a field, but is really a string-like object, which has length but no width. These theories are believed to behave as field theories, however, at energy scales well below the Planck scale.[d] In fact, these field theories contain a large number of scalar fields. Thus, the particle physics motivation for believing that scalar fields exist is quite strong.

The electromagnetic field has a potential energy density given by

$$V = \frac{1}{8\pi}\left(\mathbf{E}^2 + \mathbf{B}^2\right),$$

but the potential energy density of a scalar field has a wider range of possibilities. It is restricted by the criterion that the field theory be renormalizable (i.e. the requirement that the field theory lead ultimately to finite answers), but there are still several free parameters involved in specifying the energy density. The form of the scalar field energy density appropriate for the new inflationary universe model is shown in Fig. 7. The potential has a minimum at a value of the scalar field ϕ that is nonzero, a property that is characteristic of Higgs fields. The potential also has a flat plateau centered at $\phi = 0$, a

[d] The Planck energy E_P is given by $E_P = \sqrt{\hbar c^5/G} = 1.22 \times 10^{19}$ GeV, where G is Newton's constant, $\hbar$ is Planck's constant divided by 2π, and c is the speed of light. It is the scale above which the quantum effects of gravity are expected to become important. The Planck mass is defined by $M_P \equiv E_P/c^2 = 2.18 \times 10^{-5}$ gram.

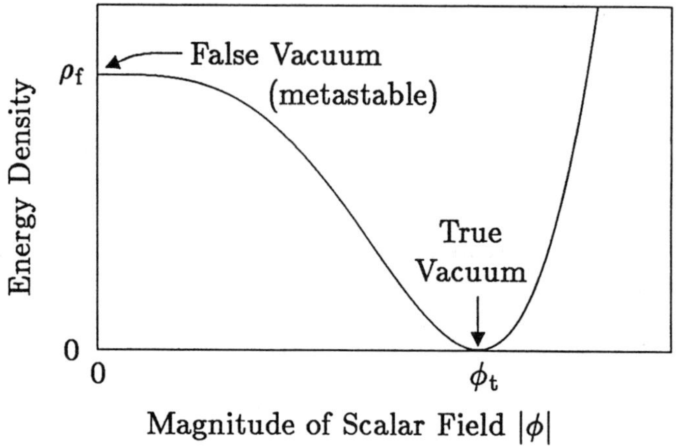

Fig. 7. The false vacuum. The graph shows the potential energy density as a function of the magnitude of a scalar field ϕ. The curve shown has a plateau at $\phi = 0$, which is the general form appropriate for the new inflationary universe model. The true vacuum is the state of lowest possible energy density, while the metastable state $\phi \approx 0$ is called the false vacuum.

feature that Higgs fields do not necessarily have. This property is necessary, however, in order for the new inflationary scenario to be possible.

A particle physicist defines the word "vacuum" to mean the state of lowest possible energy density, so it is the state in which ϕ lies at the minimum of the energy function, labeled ϕ_t on the diagram. For clarity, I will refer to the vacuum in this situation as the "true vacuum."

Notice that the energy density of the true vacuum state is shown as zero. If the energy of the vacuum were nonzero, the effects would be precisely equivalent to those of the cosmological constant introduced by Einstein in 1917. Thus the depiction of the vacuum energy density as zero on the diagram is equivalent to the statement that the cosmological constant is either zero or immeasurably small. Although the cosmological constant might be large enough to be cosmologically important, it would still be insignificant compared to ρ_f, which is estimated to be very large. In fact, the energy density of the vacuum is believed to be no larger than about $10^{-104}\ \rho_f$. The explanation for the incredible smallness of the energy density of the vacuum has been a long-standing mystery to particle physicists, although now we have at least a possible solution[34] in the context of a crude understanding of quantum gravity.

The state in which the scalar field, in some region of space, is perched at the top of the plateau of the potential energy diagram is called the "false vacuum." Note that the energy density of this state is ρ_f, and is therefore fixed by the laws of physics. (Although it seems strange that energy should be required in order for the value of the Higgs fields to be zero, particle physicists find that this property causes no inconsistencies, and is exactly what is needed to produce the symmetry breaking for which the Higgs field was introduced.) The false vacuum is of course not stable, since eventually the scalar field will no doubt evolve toward the minimum of the energy density diagram. Nonetheless, if the plateau is broad and flat enough, it will take a long time for the scalar field to roll off the hill. Thus the false vacuum is metastable, and can be very long-lived, by the time scales of the early universe.

The crucial property of the false vacuum is that the energy density is positive, but on short time scales it cannot be lowered. This explains the etymology of the name "false vacuum." Here "false" is being used to mean temporary, and "vacuum" means the state of lowest possible energy density.

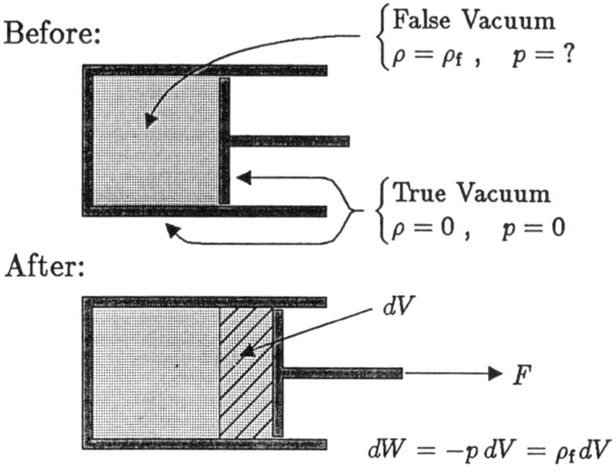

Fig. 8. Thought experiment to calculate the pressure of the false vacuum. As the piston chamber filled with false vacuum is enlarged, the energy density remains constant and the energy increases. The extra energy is supplied by the agent pulling on the piston, which must pull against the negative pressure of the false vacuum.

It is now possible to construct a simple energy-conservation argument to determine the pressure of the false vacuum. Imagine a chamber filled with false vacuum, as shown in Fig. 8. Since the energy density of the vacuum is fixed at ρ_f, the energy inside the chamber is given by $U = \rho_f V$, where V is the volume of the chamber. Now suppose the piston is pulled outward, increasing the volume by dV. Unlike any normal substance, the false vacuum will maintain a constant energy density despite the increase in volume. The increase in energy is then $dU = \rho_f dV$, which must be equal to the work done, $dW = -p dV$, where p is the pressure. Thus, the pressure of the false vacuum is given by

$$p = -\rho_f .$$

The false vacuum has a colossal, negative pressure. It is this peculiar feature of the false vacuum that is responsible for the dramatic effect that it can have on the evolution of the universe.

Assuming that the universe is undergoing homogeneous expansion as described earlier, the evolution of the scale factor $R(t)$ is given by general relativity as

$$\ddot{R} = -\frac{4\pi}{3c^2} G(\rho + 3p) R ,$$

where the overdot is used to denote a time derivative, and G is Newton's gravitational constant. In the present universe, the pressure term is a small relativistic correction. The dominant term is proportional to the energy density ρ, and the equation describes how a positive energy density produces a gravitational field that slows down the expansion of the universe. The second term shows that a positive pressure also creates an attractive gravitational field, helping to slow the expansion of the universe. However, if the universe is ever dominated by false vacuum, then during this period $p = -\rho_f$ implies that the pressure term has the opposite sign, and overcomes the gravitational attraction caused by the energy density term:

$$\ddot{R} = \frac{8\pi}{3c^2} G \rho_f R .$$

The force of gravity actually becomes repulsive, and the expansion rate of the universe is accelerated. Thus the bizarre notion of negative pressure leads to the even more bizarre effect of a gravitational force that is effectively repulsive. The general solution to this equation is

$$R(t) = c_1 e^{\chi t} + c_2 e^{-\chi t} ,$$

where c_1 and c_2 are arbitrary constants, and the exponential rate is given by

$$\chi = \sqrt{\frac{8\pi}{3c^2} G \rho_f} .$$

After some time the growing term will dominate, and the expansion becomes a pure exponential. The precise numbers here are not very well constrained, but for a typical grand unified theory one expects $\rho_f \approx (10^{15}\ \mathrm{GeV})^4/(\hbar c)^3 \approx 10^{96}$ erg/cm^3. This gives a typical value for the time constant as $1/\chi \approx 10^{-34}$ second.

While the new inflationary scenario assumes a potential energy function of the form shown in Fig. 7, Linde[35,36] has shown that such a severe restriction is not necessary. In a model known as chaotic inflation, Linde showed that inflation can work for a scalar field energy density as simple as $V(\phi) = \lambda\phi^4$, provided that one makes some assumptions about the initial conditions. He proposed that the scalar field begins in a chaotic state, so that there are some regions in which the value of ϕ is a few times larger than the Planck mass (see footnote d) M_P. These regions must exceed some minimal size, which is estimated to be several times cH^{-1}, where H denotes the Hubble constant. Then ϕ rolls down the hill of the potential energy diagram, and a straightforward calculation indicates that there is an adequate amount of inflation. The Hubble "constant" is not a constant in this case, but it is slowly varying, so the expansion can be called "quasi-exponential."

In a newer variation known as extended inflation,[37,38] the scalar field potential energy function is taken as a curve that has a local minimum in addition to the global minimum. The false vacuum is the state in which the scalar field lies in the local minimum, and inflation occurs while the universe is in this state. In this situation the false vacuum is too stable to allow a smooth ending to inflation, so an additional scalar field, which interacts directly with gravity, is introduced to bring about an end to inflation.

5. The Inflationary Universe Scenario

The inflationary universe scenario begins with a patch of the universe somehow settling into a false vacuum state. The mechanism by which this happens has no influence on the later evolution. The following three possibilities have been discussed in the literature:

1. *Supercooling from high temperatures.* This was the earliest suggestion.[29–31] If we assume that the universe began very hot, as is traditionally assumed in the standard big bang model, then as the universe cooled it presumably went through a number of phase transitions. For many types of scalar field potentials, the thermal equilibrium state at high temperatures corresponds to a scalar field that oscillates about the false vacuum value, $\phi = 0$. If such a state is rapidly cooled by the expansion of the universe, it is very natural

for the scalar field to be trapped in the false vacuum state. For the new inflationary scenario, however, this scenario has the difficulty that there is no known mechanism to achieve the desired pre-inflationary thermal equilibrium state. Calculations[39–42] show that the scalar field must be very weakly coupled in order for quantum-induced density perturbations to be sufficiently small, and consequently the scalar field would require much more than the available time to relax to thermal equilibrium. It has been shown, however, that true thermal equilibrium is not really necessary: a variety of random configurations give results that are very similar to those of thermal equilibrium.[43] In extended inflation, on the other hand, the scalar field need not be weakly coupled, so this difficulty does not arise.

2. *Tunneling from "nothing."* These ideas are of course very speculative, since they involve a theory of quantum gravity that does not actually exist. The basic idea, however, seems very plausible. If geometry is to be described by quantum theory, then the geometry of space can presumably undergo quantum transitions. One can then imagine an initial state of absolute nothingness— the absence of matter, energy, space, or time. The state of absolute nothingness can presumably undergo a quantum transition to a small universe, which then forms the initial state for an inflationary scenario. Variations of these ideas have been studied by Tyron,[44] Vilenkin,[45,46] Linde,[47–49] and Hartle & Hawking.[50]

3. *Random fluctuations in chaotic cosmology.* In Linde's[35,36] chaotic cosmology, it is assumed that the scalar field ϕ begins in a random state in which all possible values of ϕ occur. Inflation then takes place in those regions that have appropriate values of ϕ, and these inflated regions dominate the universe at later times. In these models it is not necessary for the scalar field potential energy function $V(\phi)$ to have a plateau, but as in other models it must be very flat (i.e. weakly self-coupled) in order to minimize the density perturbations that result from quantum fluctuations.

Regardless of which of the above mechanisms is assumed, one expects that the scalar field just before inflation would be approximately uniform over distances equal to the age of the universe times the speed of light. For typical grand unified theory parameters, this gives uniformity on a length scale of about 10^{-24} cm. The inflationary model posits that the entire observed universe evolved from a region of this size or smaller.

The patch then expands exponentially due to the gravitational repulsion of the false vacuum. In the inflationary model, it is the force of this repulsion that provides the "bang" of the big bang, creating the cosmic expansion which we know as Hubble's law. (The standard big bang theory, by contrast, is really only a theory of the aftermath of a bang — the cosmic expansion is not explained, but instead is incorporated into the postulated initial conditions.) In order to achieve the goals of inflation, we must assume that this exponential expansion results in an expansion factor $\gtrsim 10^{25}$. For typical grand unified theory numbers, the universe would double in size during each interval of about 10^{-34} second, so this enormous expansion requires only about 10^{-32} second of inflation. The actual amount of inflation, however, was very likely much more than this required minimum.

During the inflationary period, the density of any particles that may have been present before inflation is diluted so much that it becomes completely negligible. The regions within which the scalar field is approximately uniform are stretched by the expansion factor to become at least about 10 cm across.[e] If the duration of inflation is more than the minimal value, which seems quite likely, then these regions could be many orders of magnitude larger. There appears to be no upper limit to the amount of inflation that may have taken place.

The false vacuum is not stable, so it eventually decays. If the decay occurs by the usual process of bubble nucleation,[51,52] similar to the way water boils, then the randomness of the bubbles would produce gross inhomogeneities in the mass density.[53,54] This problem is avoided in the new inflationary scenario[30,31] by introducing a scalar field potential energy function with a flat plateau, as was shown in Fig. 7. This leads to a "slow-rollover" phase transition, in which quantum fluctuations destabilize the false vacuum, starting the scalar field to roll down the hill of the potential energy diagram. These fluctuations are initially correlated over only a microscopic region, but the additional inflation that takes place during the rolling can stretch such a region to be large enough to easily encompass the observed universe.

[e]The alert reader might notice that these numbers imply that the region expands much faster than the speed of light. Even in the standard big bang model without inflation, the size of the universe, or the observed part of the universe, increases faster than the speed of light. Although this violates the premises of special relativity, it is completely acceptable in the context of general relativity. The principle that nothing can travel faster than light is valid in general relativity, but it has to be defined more carefully than in special relativity. It is accurate to say that if any particle has a race with a light beam, the light beam will *always* win. The complication, however, is that in general relativity space is not rigid, but instead is a plastic medium that can bend and twist and stretch. So, if two points are separated from each other, the distance between them can be increased by the stretching of the space between them. There is nothing in general relativity that places any limit on the speed with which such stretching can take place.

When the phase transition takes place, the energy that has been stored in the false vacuum is released in the form of new particles. (In the language of thermodynamics, this energy is the "latent heat" of the phase transition.) These new particles rapidly come to thermal equilibrium, resulting in a temperature with $kT \approx 10^{14}$ GeV. At this point the scenario rejoins the standard cosmological model. The presently observed region of the universe has a radius of about 10 cm at this time; the normal Hubble expansion of standard cosmology stretches this radius to about 10^{10} light-years by today.

The excess of matter over antimatter is produced[55] after inflation, since any excess that may have been present before inflation would have been diluted away by the enormous expansion. Thus, inflationary cosmology requires an underlying particle theory in which baryon number, the number of protons and neutrons minus the number of antiprotons and antineutrons, is not exactly conserved. Grand unified theories fulfill this requirement, and it now seems clear that baryon number violation occurs at high temperatures even in the well-established theory of electromagnetic and weak interactions.[56–59] At the end of inflation, the universe would have been uniformly filled with a hot gas of particles, exactly as had been postulated as the initial condition for the standard big bang theory. The inflationary model merges with the standard big bang theory, so the two models agree in their description of the evolution of the universe from this time onward.

It is a dramatic feature of inflationary models that essentially all of the matter, energy, and entropy of the observed universe is produced by the expansion and subsequent decay of the false vacuum. (I used the qualifier "essentially" to acknowledge the fact that a small patch of false vacuum is necessary to start inflation.) This seems strange, because it sounds like an unmistakable violation of the principle of energy conservation. How could it be possible that all the energy in the universe was produced as the system evolved?

In fact, the inflationary universe model is consistent with all the known laws of physics, including the conservation of energy. The loophole in the conservation of energy argument is due to the peculiar nature of gravitational energy. Using either general relativity or Newtonian gravity, one finds that *negative* energy is stored in a gravitational field. A simple way to demonstrate this fact is to imagine a spherical shell of mass, as shown in Fig. 9(a). The gravitational field outside the shell points radially inward, and has the same strength as it would if all of the mass were concentrated at the center. In the spherical cavity inside the shell, however, the gravitational field caused by the different parts of the shell cancel out exactly, so the gravitational field is zero. The shell itself will experience an inward gravitational force, since the

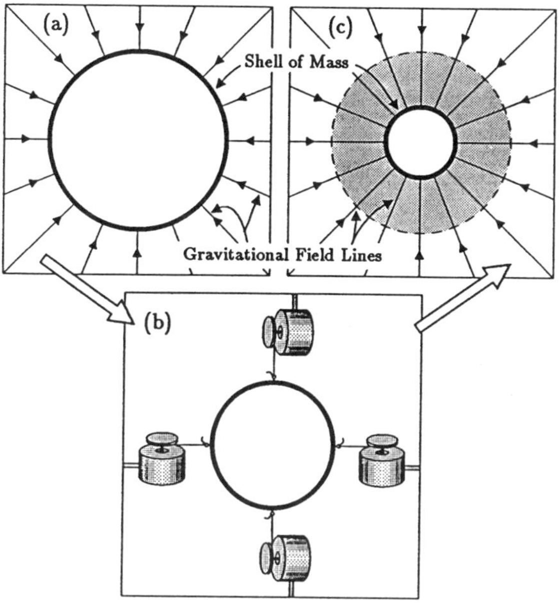

Fig. 9. Thought experiment to understand the energy of gravity. Figure (a) shows a hollow spherical shell of mass, and the gravitational field lines that it produces. There is a force on each piece of the shell, pulling inward. Figure (b) shows how energy can be extracted as the shell is allowed to uniformly contract. Each piece of the shell is tied by a rope to an electrical generator, producing power as the piece is "lowered" toward its final position. Figure (c) shows the final configuration, which includes a gravitational field in the shaded region where no field existed before. Thus, the creation of the gravitational field is associated with the release of energy.

gravitational field within the matter of the shell is nonzero, varying smoothly between the outside value and zero. Now imagine what would happen if the shell were allowed to uniformly contract. Assume that the matter from which the shell is constructed is soft and compressible, so it offers no mechanical resistance to the contraction. One can imagine, for example, extracting energy by tying ropes to each piece of the shell, as is illustrated in Fig. 9(b). These ropes can be used to drive electric generators as each piece is lowered to its new position. After the new radius is attained, the picture will look like Fig. 9(c). The dotted circle indicates the former radius of the shell, and outside of the dotted circle the gravitational field is identical to that in Fig. 9(a). Inside the shell in its new position, the gravitational field remains zero. However, in the shaded region between the original and new positions of the shell, a gravitational field now exists where no field had existed before. Thus, the net effect of this operation is to extract energy, and to create a new region of gravitational field. Thus, energy is *released* when a gravitational field is created. If the absence of a gravitational field corresponds to zero energy, then any nonzero field strength must correspond to a negative energy. In most physical processes the exchange of gravitational energy is much smaller than the rest energy (mc^2) of the particles involved, but cosmologically the total gravitational energy can be large enough to cancel the rest energy.

In the inflationary model, the energy stored in the false vacuum becomes larger and larger as the universe inflates, and is then released when the phase transition takes place at the end of the inflationary period. At the same time, however, the energy stored in the cosmic gravitational field — the field by which everything in the universe is attracting everything else — becomes more and more negative. The total energy of the system is conserved, remaining constant at a value which is either zero or very small.

Thus, inflation allows the entire observed universe to develop from almost nothing. The initial volume must have a minimum radius of approximately cH^{-1}, where H is the Hubble constant at the time of inflation, which for typical grand unified theory parameters implies a minimum mass of only about 10 kilograms. (Even this initial mass, however, could conceivably have been balanced by an equal contribution of negative energy in the gravitational field.) Thus, if the inflationary model is correct, it is fair to say that the universe is the ultimate free lunch.

6. Answers to Three Questions

Having described the foundations of the inflationary universe model, I can now explain how the three questions of Table 1 can be resolved.

First, I will discuss Question #1, concerning the large-scale homogeneity of the universe. Recall that in the standard big bang theory, the large-scale homogeneity cannot be explained because the universe did not have enough time to come to a uniform temperature.

Consider now the evolution of the observed region of the universe, which has a radius today of roughly 10 billion light-years. Imagine following this region backward in time, using the inflationary model. Start by following it back to the instant just after the inflationary period. The evolution during this interval — from the present back to just after inflation — is the same as in standard cosmology. Continuing further backward we come to the enormous expansion of the inflationary era. Going backward in time, we see the inflationary period as an enormous contraction. Thus, just before inflation we find that the region was incredibly small. In fact, the region was more than a billion times smaller than the size of a proton. (Note that I am *not* asserting that the universe as a whole was this small. The inflationary model makes no statement about the size of the universe as a whole, which could in fact even be infinite.)

While the region was incredibly small before inflation, there was plenty of time for it to have come to a uniform temperature by the same kind of mundane processes that cause a glass of ice water to come to a uniform temperature. So, in the inflationary model, the uniform temperature was established before inflation took place, in a very, very small region. The process of inflation then magnified this very small region to become large enough to encompass the entire observed universe. Thus, the sources of the microwave background radiation arriving today from all directions in the sky were once in close contact; they had time to reach a common temperature before the inflationary era began.[f]

The inflationary model also provides a simple resolution for Question #2, the issue of the mass density. Recall that the ratio of the actual mass density to the critical density is called Ω, and that the problem arose because the condition $\Omega = 1$ is unstable: Ω is always driven away from one as the universe evolves, making it difficult to understand how its value today can be in the vicinity of one.

During the inflationary era, however, the peculiar nature of the false vacuum state results in some important sign changes in the equations that describe the evolution of the universe. During this period, as we have seen, the force of gravity acts to accelerate the expansion of the universe rather than to retard it. It turns out that the equation governing the evolution of Ω also has a crucial change of sign: during the inflationary period the universe is driven very quickly and very powerfully *toward* a critical mass density.

In other words, a very short period of inflation can drive the value of Ω very accurately to one, almost no matter where it starts out. There is no longer any need to assume that the initial value of Ω was incredibly close to one.

[f]Since all distances go to zero at the instant of the big bang singularity, with or without inflation, the astute reader will ask why causal contact and thermal equilibrium cannot be achieved at this time. The reason is that the expansion velocities approach infinity at the singularity. (If you are puzzled by a speed greater than that of light, see footnote e.) So, even though the distance between two particles of matter was zero at the singularity, the infinite velocity of separation implies that no causal contact could have taken place. One determines the question of causal contact by tracing the trajectory of a light ray that leaves one particle immediately after the big bang singularity. One finds that the light ray reaches the other particle much later in the history of the universe, as was illustrated in Fig. 6.

Furthermore, there is a prediction that comes out of this. The mechanism that drives Ω to one almost always overshoots, which means that even today the mass density should be equal to the critical mass density to a high degree of accuracy. More precisely, the model predicts that the value of Ω today should equal one to an accuracy of about one part in 100,000. (The deviations from one are caused by quantum effects, which I will talk about shortly.) Thus, the determination of the mass density of the universe would be a very important test of the inflationary model.[g]

Unfortunately, it is very difficult to estimate the mass density of the universe reliably. Part of the reason is that most of the mass in the universe is in the form of "dark matter," matter that is totally unobserved except for its gravitational effects on other forms of matter. Since we do not even known what the dark matter is, it is very difficult to estimate how much exists. If one counts only the matter that is directly visible in the form of stars and galaxies, one finds a density only 1–2% of the critical value. However, measurements of the velocities of stars within galaxies and of galaxies within clusters of galaxies indicate that all of these velocities are surprisingly high. Since these structures do not fly apart, we infer that they must be held together by the gravitational field created by a rather large amount of dark matter. When this dark matter is included, the estimates[60] of Ω rise to something like 0.1 to 0.3. Dark matter, however, can be detected only when there is visible matter present for it to influence, so there remains the possibility that there is much more dark matter than has so far been found. In fact, studies of the velocities and densities of galaxies on a very large scale[61,62] (within a radius of about 600 million light-years) strongly suggest an Ω in the vicinity of 1.

Finally, then, I come to the last of the three questions, concerning the origin of the primordial density perturbations in the universe.

The generation of density perturbations in the new inflationary universe was addressed in the summer of 1982 at the Nuffield Workshop on the Very Early Universe, held at Cambridge University. A number of theorists were working on this problem, including Steinhardt, James M. Bardeen of the University of Washington, Stephen W. Hawking of Cambridge University, Michael S. Turner of the University of Chicago, A. A. Starobinsky of the L. D. Landau Institute of Theoretical Physics in Moscow, and me. (My work was done

[g]In the text I have followed the common assumption that Einstein's cosmological constant Λ is either zero or negligible. Otherwise the prediction becomes $\Omega + (\Lambda/3H^2) = 1$. The cosmological constant can be interpreted as a nonzero energy density in the vacuum, so $\Omega + (\Lambda/3H^2) = 1$ implies that the sum of the matter and vacuum contributions to the energy density is equal to the critical density.

in collaboration with So-Young Pi of Boston University, although she did not attend the workshop.) We found that the new inflationary model, unlike any previous cosmological model, leads to a definite prediction for the spectrum of perturbations. Basically, the process of inflation first smooths out any primordial inhomogeneities that may have been present in the initial conditions. For example, any particles that may have been present before inflation are diluted to a negligible density. In addition, the primordial universe may have contained inhomogeneities in the gravitational field, which is described in general relativity in terms of bends and folds in the structure of spacetime. Inflation, however, stretches these bends and folds until they become imperceptible, just as the curvature of the surface of the Earth is imperceptible in our everyday lives. For a while, we were worried that inflation would give us a totally smooth universe, which would be obviously incompatible with observation. It was pointed out, however, I believe first by Hawking, that the situation might be saved by the application of quantum theory.

A very important property of quantum physics is that nothing is determined exactly — everything is probabilistic. Physicists are, of course, accustomed to the idea that quantum theory, with its probabilistic predictions, is essential to describe phenomena on the scale of atoms and molecules. On the scale of galaxies or clusters of galaxies, however, there is usually no need to consider the effects of quantum theory. But inflationary cosmology implies that for a short period the scales of distance increased very rapidly with time. Thus, the quantum effects which occurred on very small, particle-physics length scales were quickly stretched to the scales of galaxies and clusters of galaxies by the process of inflation.

Therefore, even though inflation with the rules of classical physics would predict a completely uniform mass density, the inherent probabilistic nature of quantum theory gives rise to small perturbations about the classically predicted value. The spectrum of these perturbations was first calculated during the exciting three-week period of the Nuffield Workshop.[39–42] After much disagreement and discussion, the various working groups came to an agreement on the answer. I will describe these results in two parts.

First of all, we calculated the shape of the spectrum of the perturbations. The concept of a spectrum of density perturbations may seem a bit foreign, but the analogy of sound waves is very close. People familiar with acoustics understand that no matter how complicated a sound wave is, it is always possible to carry out a Fourier expansion, breaking it up into components that each have a standard (sine wave) wave form and a well-defined wavelength. The spectrum of the sound wave is specified by the strength of each of these components. In discussing density perturbations in the universe, it is similarly useful to define a spectrum by breaking up the perturbations into sine waves of well-defined wavelength.

For the inflationary model, we found that the predicted shape for the spectrum of density perturbations is essentially scale-invariant; that is, the strength of the perturbations is approximately equal on all wavelengths of astrophysical significance. While the precise shape of the spectrum depends on the details of the underlying particle physics, the approximate scale invariance holds in almost all cases. The scale-invariant spectrum is in agreement with a phenomenological model for galaxy formation proposed in the early 1970's by Edward R. Harrison[63] of the University of Massachusetts at Amherst and Yakov B. Zel'dovich[64] of the Institute of Physical Problems in Moscow, working independently.

Unfortunately, there is still no way of inferring the precise form of the primordial spectrum from observations, since one cannot reliably calculate how the universe evolved from the early period to the present. Such a calculation is very difficult in any case, and it is further complicated by the uncertainties about the nature of the dark matter. Nonetheless, the scale-invariant spectrum appears to be at least approximately what is needed to explain the evolution of galaxies, and thus this prediction of the inflationary model appears so far to be successful.

Furthermore, the recent results from the COBE satellite on nonuniformities in the cosmic background radiation provide a direct test of the spectrum as predicted by inflation.[h] The satellite measured δT, the deviation of the cosmic background radiation temperature from the mean, as a function of angular position (θ, ϕ). The COBE team then computed the correlation function $C(\alpha) \equiv \langle \delta T_1 \delta T_2 \rangle$, the average value of the product of δT at two different points,[i] averaged over all pairs of points with a given angular separation α. The plot of the COBE correlation function[18] is shown as Fig. 10. The shaded region indicates the range expected for a scale-invariant spectrum, normalized to the best fit of the data. Note that the prediction gives a range of expected values — it is a quantum-mechanical prediction, so only a probability distribution can be inferred. (Precisely, the shaded region indicates the mean plus or minus one standard deviation.) The graph shows a beautiful agreement with the predictions of inflation. Quantitatively, the spectrum can be characterized

[h] The COBE data was not available when this talk was given, but the result is so important that I could not possibly omit it from the written version.

[i] In evaluating δT, the effect of the earth's motion through the background radiation was subtracted by a best-fit procedure.

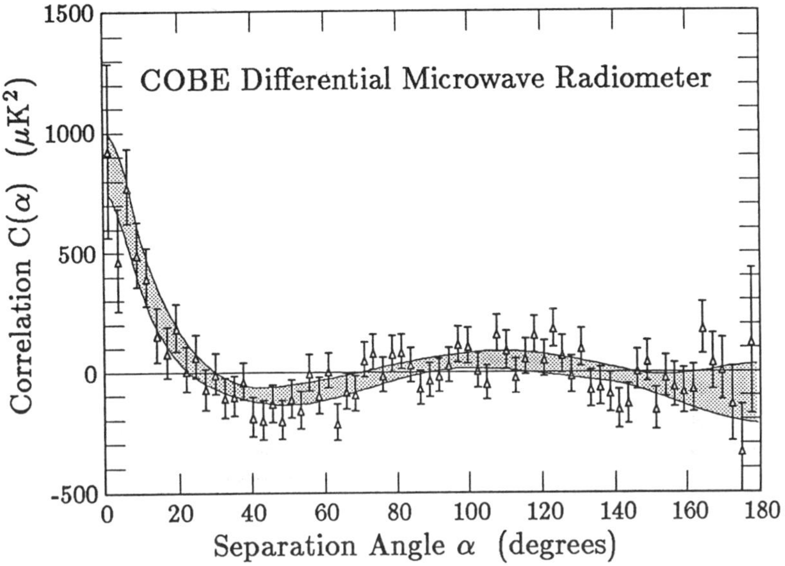

Fig. 10. The COBE anisotropy results. The graph shows the correlation function $C(\alpha) \equiv \langle \delta T_1 \delta T_2 \rangle$, where δT denotes the deviation from the average of the cosmic background radiation temperature in a given direction, and the brackets denote an average over all pairs of directions that are separated from each other by an angle α. The shaded region denotes the range of values predicted by the inflationary model.

by a spectral index n, which is 1 for a scale-invariant spectrum. (The spectral power is proportional to k^n, where k is the inverse of the wavelength.) Analysis of the COBE data implies that $n = 1.1 \pm 0.5$.

The angular resolution of the COBE satellite is about 7°, so it is impossible for COBE to see the variations of the cosmic microwave background on small angular scales. Thus, the length scales probed by the COBE data are much longer than the length scales that are directly relevant to structure formation. It is possible, therefore, to improve the test of scale invariance by combining the COBE data with estimates of the spectrum at shorter wavelengths derived from theories of structure formation. When this is done,[65] one obtains a baseline spanning three decades in length scale, with a best-fit value $n = 1 \pm 0.23$. The uncertainties are smaller but still significant, and the agreement with the prediction of inflation remains excellent.

Galaxy formation is currently a very active subject of research, so a better determination of the spectrum of primordial density perturbations may be developed. Such a result would provide a more discriminating test of the inflationary model.

The predicted magnitude of the density perturbations was also calculated by the group at the Nuffield Workshop, but the implications of these results were much less clear. It was found that the predicted magnitude, unlike the shape of the spectrum, is very sensitive to the details of the underlying particle theory. At the time, the minimal SU(5) theory — the first and simplest of the grand unified theories — was strongly favored by anybody working on the subject. We were therefore very disappointed when we found that the minimal SU(5) theory leads to density perturbations with a magnitude 100,000 times larger than what is desired for the evolution of galaxies. Thus, there was a serious incompatibility between the inflationary model and the simplest of the grand unified theories.

With the passage of time, however, the credibility of the minimal SU(5) grand unified theory has diminished. The minimal SU(5) theory makes a rather definite prediction for the lifetime of a proton, and a variety of experiments have been set up to test this prediction by looking for proton decay. So far, no such decays have been observed, and the experiments have pushed the limit on the proton lifetime to the point where the minimal SU(5) theory is now excluded.

With the exclusion of the minimal SU(5) theory, a wide range of grand unified theories become plausible. All of the allowed theories seem a bit complicated, so apparently we will need some kind of new understanding to choose which — if any — is correct.

A variety of grand unified theories that predict an acceptable magnitude for both the proton lifetime and the density perturbations have been constructed. Thus, while the inflationary model cannot be credited with correctly predicting the magnitude of the perturbations, it also cannot be criticized for making a wrong prediction. The inflationary model provides at least a framework for calculating the magnitude of the density perturbations. If sometime in the future the correct particle theory and the values of its free parameters somehow become known, it will then be possible to make a real theoretical prediction for the magnitude of the perturbations.

A common feature of those models leading to acceptable density perturbations is the abandonment of the original hope that inflation can be driven by the Higgs fields that play an integral role in the grand unified theories. It appears that any Higgs field that interacts strongly enough to break the grand unified symmetry leads to density perturbations with a magnitude that is far too large. Thus it must be assumed that the underlying particle theory contains a new field — a field which strongly resembles the Higgs fields in its qualitative properties, but which interacts much more weakly. In particular, if the scalar field energy density contains a nonlinear[j] term $\lambda\phi^4$, then λ must be less than about 10^{-13}.

Unfortunately, all of the known theories that give acceptable predictions for the magnitude of the density perturbations look a little contrived. Well, to be honest the theories *were* contrived — with the goal of getting the density perturbations to come out right. The need for this contrivance can certainly be used as an argument against the inflationary model; but, in my opinion, this argument is considerably weaker than the arguments in favor of inflation. Even if we ignore cosmology, any known particle theory that is consistent with the known properties of particle physics appears to be rather contrived. In particular, there are a number of very small dimensionless quantities in nature that are not understood: the ratio of the weak energy scale to the Planck scale is about 10^{-17}, and the Yukawa coupling constant that describes the interaction of the electron and the Higgs field is about 10^{-6}. Clearly there are some fundamental principles at work here that we are still missing.

Finally, I want to mention that quantum effects during the inflationary era are not the only potential source of primordial density perturbations. One alternative possibility, which I will not discuss in detail, is that the seeds for galaxy formation may have been objects called "cosmic strings" (not related to "superstrings"). These strings are predicted by some (but not all) grand unified theories, and they would have formed in a random pattern during the grand unified theory phase transition. As their name suggests, strings are very thin, spaghetti-like objects that can form infinite curves or closed loops of astrophysical size. With a thickness of about 10^{-29} centimeter, a cosmic string has a mass of about 10^{22} grams for each centimeter of length. (In astronomical terms that is equal to about 10^7 solar masses per light-year.) In most theories, the density of these strings would be diluted to negligibility by the process of inflation. However, it is possible to construct theories in which the strings survive by forming either after inflation or at the very end of it. Cosmic strings produce a nearly scale-invariant spectrum of density perturbations very similar to inflation, so it is difficult to distinguish observationally. Models of this type

[j] The connection between nonlinear terms and particle interactions is well known to particle physicists, but has not yet been discussed in this paper. To understand the connection, note that classical wave packets described by linear equations can pass through each other as a linear superposition, with no interaction. Wave packets interact therefore only through the nonlinear terms in the equations of motion, and this connection carries over to the description of particles in the quantized theory.

still make use of inflation to answer Questions #1 and #2 of Table 1, and also to smooth out any small-scale inhomogeneities which may have been present in the initial conditions.

7. The Eternal Nature of Inflation

A fascinating feature of inflation, which in my opinion is also important in evaluating the plausibility of inflation, is the fact that inflation is eternal—if inflation ever begins, then it will never stop.[66–70]

To understand the endlessness of inflation, one first notices that the decay of the false vacuum, like the decay of many other unstable systems, is an exponential process. S.-Y. Pi and I[71] have verified the exponential decay law in a simplified but exactly soluble model of a slow-rollover phase transition, in which the potential energy function is taken as $V(\phi) = -\frac{1}{2}\mu^2\phi^2$. For the case of chaotic inflation, on the other hand, one might think that the scalar field would roll inexorably down the hill in the potential energy diagram, completing the decay in a finite time. Linde[68] has shown, however, that if the scalar field starts at a sufficiently high value, then it can be sustained by quantum fluctuations, with again an exponential decay law. As long as the false vacuum endures it drives an exponential expansion, and for reasonable parameters the rate of expansion is much faster than the rate of decay. Thus, even though the false vacuum is decaying, the total volume of the false vacuum region actually increases with time.

As time goes on, pieces of the false vacuum region are constantly undergoing decay. As each piece decays, it releases energy and thereby sets in motion a hot big bang universe. Other regions of false vacuum, however, continue to exponentially expand, so the false vacuum never disappears.

The infinity of universes produced in this way rapidly become causally disconnected, so there is no way for us to verify, even in principle, that the other universes exist. Nonetheless, I feel that the eternal character of inflation makes it a more plausible theory. In the absence of this feature, there is some difficulty in deciding whether the initial conditions required for inflation are sufficiently plausible. Since there is no established theory of initial conditions, questions of this sort can easily lead to inconclusive answers. Given the endlessness of inflation, however, the question becomes much less significant. Just as most of us accept the claim that complicated DNA molecules originated through random processes sometime during the history of the earth, we can also accept the claim that a region of false vacuum originated through random processes sometime during the history of spacetime. Just as primitive DNA molecules increased their abundance by replication, one patch of false vacuum would inflate to produce an infinity of universes, one of which could be the universe in which we live.

8. Conclusion

In summary, the inflationary universe model has been very successful in describing the broad, qualitative properties of the universe. In particular, the model provides very attractive answers to the three questions discussed in this lecture. While the model must be treated as speculative, I nonetheless feel that in its broad outline the concept of an inflationary universe is essentially correct.

The inflationary model makes two primary predictions — it predicts the mass density of the universe, and also the shape of the spectrum of primordial density perturbations. The prediction of the mass density is extremely firm, while the prediction of the perturbation spectrum depends weakly on the underlying particle physics. While neither of these predictions is straightforward to check, it seems likely that significant progress will be made in the foreseeable future. The COBE results have already provided a significant check on the spectrum of primordial density fluctuations, and the results are completely in accord with inflation.

Even if the inflationary model is correct, however, I must still emphasize that nothing discussed here is a completed project. The inflationary model is not a detailed theory — it is really just an outline for a theory — Michael Turner has called it the "inflationary paradigm." To fill in the details, we will need to know much more about the properties of particle physics at the energy scale of grand unified theories — and perhaps beyond. And to understand the evolution of the universe to the present time, we will of course have to identify the dark matter and solve a host of other astrophysical problems.

It looks to me as if the fields of particle physics and cosmology will remain closely linked for some years to come, as physicists and astronomers continue their efforts to understand the structure and history of the universe.

Oskar Klein was very much concerned with the question of finding the driving force behind the expansion of the universe, and he wrote a paper with Hannes Alfvén[72] proposing that the expansion was driven by matter–antimatter annihilation. This idea has not been very successful, as it is very difficult to understand how the extreme uniformity of the cosmic background radiation can emerge from any kind of ordinary explosion. It is always hard to judge another person's taste, but I suspect that Oskar Klein would be very happy to know that a plausible driving force behind the big bang can be found, coming directly from the Oklein–Gordon equation.

References

1. *NYNEX White Pages: Boston, Including Brookline, Cambridge, Somerville*, 1992 (NYNEX Information Resources Co., 1991).
2. For the reader who would like a more detailed but still nontechnical treatment of these topics, I would recommend the discussion of the standard big bang model by Steven Weinberg, *The First Three Minutes* (Basic Books, New York, 1977, 188 pp.), and the discussions of the more recent developments by John Gribbin, *In Search of the Big Bang: Quantum Physics and Cosmology* (Heinemann, London, 1986, 413 pp.), Lawrence M. Krauss, *The Fifth Essence: The Search for Dark Matter in the Universe* (Basic Books, New York, 1989, 342 pp.), and Michael Riordan and David N. Schramm, *The Shadows of Creation: Dark Matter and the Structure of the Universe* (W. H. Freeman and Company, New York, 1991, 277 pp.). At a more technical level, an excellent description of these developments can be found in Edward W. Kolb and Michael S. Turner, *The Early Universe* (Addison-Wesley, Redwood City, Calif., 1990, 547 pp.).
3. The history of Einstein's contributions is very well described in Abraham Pais, *"Subtle Is the Lord ... ": The Science and the Life of Albert Einstein* (Oxford University Press, New York, 1982, 552 pp.).
4. A. Einstein, "Die Grundlage der allgemeinen Relativitätstheorie," *Ann. Phys.* **49**, 769–822. English translation in Ref. 5.
5. H. A. Lorentz, A. Einstein, H. Minkowski, and H. Weyl, *The Principle of Relativity: A Collection of Original Memoirs on the Special and General Theory of Relativity.* Translated by W. Perrett and G. B. Jeffery. (Dover, New York, 1952, 216 pp.).
6. A. Einstein, "Kosmologische Betrachtungen zur allgemeinen Relativitätstheorie" ("Cosmological considerations on the general theory of relativity"), *Sitzungsberichte der Preussischen Akademie der Wissenschaften* **1917**, 142–152 (1917). English translations appear in Refs. 5 and 7.
7. J. Bernstein and G. Feinberg, eds., *Cosmological Constants: Papers in Modern Cosmology* (Columbia University Press, New York, 1987, 328 pp.).
8. For an informative discussion of Newton's involvement with this problem, see Edward Harrison, "Newton and the infinite universe," *Phys. Today*, February 1986, p. 24.
9. E. Hubble, "A relation between distance and radial velocity among extra-galactic nebulae," *Proc. Natl. Acad. Sci.* **15**, 168–173 (1929). (Reprinted in Ref. 7.)
10. W. Baade, "Report of the meeting of the commission on extragalactic nebulae, Eight General Assembly (Rome, September 4–13, 1952)," *Transactions of the International Astronomical Union* **8**, 397–399 (1952).
11. A. Sandage, "Current problems in the extragalactic distance scale," *Astrophys. J.* **127**, 513–526 (1958).
12. A. A. Penzias and R. W. Wilson, "A measurement of excess antenna temperature at 4080 Mc/s," *Astrophys. J.* **142**, 414–419 (1965).

13. J. Silk, *The Big Bang* (W. H. Freeman and Company, New York, 1980), p. 78.
14. T. Matsumoto, S. Hayakawa, H. Matsuo, H. Murakami, S. Sato, A. E. Lange, and P. L. Richards, "The submillimeter spectrum of the cosmic background radiation," *Astrophys. J.* **329**, 567–571 (1988).
15. J. C. Mather *et al.*, "A preliminary measurement of the cosmic microwave background spectrum by the Cosmic Background Explorer (COBE) satellite," *Astrophys. J. Lett.* **354**, L37–40 (1990).
16. Figure adapted from D. N. Schramm, "The early universe and high-energy physics," *Phys. Today*, April 1983, pp. 27–33.
17. J. Yang, M. S. Turner, G. Steigman, D. N. Schramm, and K. A. Olive, "Primordial nucleosynthesis: a critical comparison of theory and observation," *Astrophys. J.* **281**, 493–511 (1984).
18. G. F. Smoot *et al.*, "Structure in the COBE differential microwave radiometer first year maps," *Astrophys. J. Lett.* **396**, L1–L6 (1992).
19. R. H. Dicke and P. J. E. Peebles, "The big bang cosmology — enigmas and nostrums," in *General Relativity: An Einstein Centenary Survey*, eds. S. W. Hawking and W. Israel (Cambridge University Press, Cambridge, 1979), pp. 504–517.
20. For the reader who would like a more detailed but still nontechnical treatment, I would recommend Refs. 21 and 22. For the reader looking for a more technical review, I would recommend any of Refs. 23–28.
21. Edward P. Tryon, "Cosmic Inflation," in *The Encyclopedia of Physical Science and Technology* (Academic, New York, 1987), Vol. 3, pp. 709–743.
22. Alan H. Guth and Paul J. Steinhardt, "The inflationary universe," in *The New Physics*, ed. Paul Davies (Cambridge University Press, Cambridge, 1988), pp. 34–60.
23. L. F. Abbott and S. Y. Pi, eds., *Inflationary Cosmology* (World Scientific, Singapore, 1986, 697 pp.).
24. Andrei Linde, *Particle Physics and Inflationary Cosmology* (Harwood, Chur, Switzerland, 1990, 362 pp.).
25. Michael S. Turner, "Cosmology and particle physics," in *Architecture of Fundamental Interactions at Short Distances*, eds. P. Ramond and R. Stora (North-Holland, Amsterdam, 1987), pp. 513–680.
26. Robert H. Brandenberger, "Quantum field theory methods and inflationary universe models," *Rev. Mod. Phys.* **57**, 1–60 (1985).
27. Paul J. Steinhardt, "Inflationary cosmology," in *High Energy Physics, 1985, Proceedings of the Yale Theoretical Advanced Study Institute*, eds. M. J. Bowick and F. Gürsey (World Scientific, 1986), Vol. 2, pp. 567–617.
28. S. K. Blau and A. H. Guth, "Inflationary cosmology," in *300 Years of Gravitation*, eds. S. W. Hawking and W. Israel (Cambridge University Press, Cambridge, 1987), pp. 524–603.
29. A. H. Guth, "Inflationary universe: a possible solution to the horizon and flatness problems," *Phys. Rev.* **D23**, 347–356 (1981).
30. A. D. Linde, "A new inflationary universe scenario: a possible solution of the horizon, flatness, homogeneity, isotropy and primordial monopole problems," *Phys. Lett.* **108B**, 389–393 (1982).

31. A. Albrecht and P. J. Steinhardt, "Cosmology for grand unified theories with radiatively induced symmetry breaking," *Phys. Rev. Lett.* **48**, 1220–1223 (1982).
32. P. Jordan and O. Klein, "Zum Mehrkörperproblem der Quantentheorie" ("On the many-body problem of quantum theory"), *Z. Phys.* **45**, 751–765 (1927).
33. O. Klein, "Elektrodynamik und Wellenmechanik vom Standpunkt des Korrespondenzprinzips" ("Electrodynamics and wave mechanics from the standpoint of the correspondence principle"), *Z. Phys.* **41**, 407–442 (1927).
34. Sidney Coleman, "Why there is nothing rather than something: a theory of the cosmological constant," *Nucl. Phys.* **B310**, 643 (1988).
35. A. D. Linde, "Chaotic inflating universe," *Pis'ma Zh. Eksp. Teor. Fiz.* **38**, 149–151 (1983). [English translation: *JETP Lett.* **38**, 176–179 (1983)].
36. A. D. Linde, "Chaotic inflation," *Phys. Lett.* **129B**, 177–181 (1983).
37. D. La and P. J. Steinhardt, "Extended inflationary cosmology," *Phys. Rev. Lett.* **62**, 376–378 (1989).
38. P. J. Steinhardt and F. S. Accetta, "Hyperextended inflation," *Phys. Rev. Lett.* **64**, 2740–2743 (1990).
39. A. A. Starobinsky, "Dynamics of phase transition in the new inflationary universe scenario and generation of perturbations," *Phys. Lett.* **117B**, 175–178 (1982).
40. A. H. Guth and S.-Y. Pi, "Fluctuations in the new inflationary universe," *Phys. Rev. Lett.* **49**, 1110–1113 (1982).
41. S. W. Hawking, "The development of irregularities in a single bubble inflationary universe," *Phys. Lett.* **115B**, 295–297 (1982).
42. J. M. Bardeen, P. J. Steinhardt, and M. S. Turner, "Spontaneous creation of almost scale-free density perturbations in an inflationary universe," *Phys. Rev.* **D28**, 679–693 (1983).
43. R. Brandenberger, H. Feldman and J. Kung, "Initial conditions for chaotic inflation," *Physica Scripta* **T36**, 64–69 (1991). Also published in *The Birth and Early Evolution of Our Universe*, eds. J. S. Nilsson, B. Gustafsson, and B.-S. Skagerstam (World Scientific, Singapore, 1991), pp. 64–69.
44. E. P. Tyron, "Is the universe a vacuum fluctuation?" *Nature* **246**, 396–397 (1973).
45. A. Vilenkin, "Creation of universes from nothing," *Phys. Lett.* **117B**, 25–28 (1982).
46. A. Vilenkin, "Quantum origin of the universe," *Nucl. Phys.* **B252**, 141–151 (1985).
47. A. D. Linde, "The new inflationary universe scenario," in *The Very Early Universe: Proceedings of the Nuffield Workshop*, eds. G. W. Gibbons, S. W. Hawking, and S. T. C. Siklos (Cambridge University Press, Cambridge, England, 1983), pp. 205–249.
48. A. D. Linde, "Quantum creation of the inflationary universe," *Lett. Nuovo Cimento* **39**, 401–405 (1984).
49. A. D. Linde, "Inflation and quantum cosmology," in *300 Years of Gravitation*, eds. S. W. Hawking and W. Israel (Cambridge University Press, Cambridge, 1987), pp. 604–630.
50. J. B. Hartle and S. W. Hawking, "Wave function of the universe," *Phys. Rev.* **D28**, 2960–2975 (1983).

51. S. Coleman, "The fate of the false vacuum, 1: semiclassical theory," *Phys. Rev.* **D15**, 2929–2936 (1977) [Errata: **16**, 1248 (1977).]
52. C. G. Callan and S. Coleman, "The fate of the false vacuum, 2: first quantum corrections," *Phys. Rev.* **D16**, 1762–1768 (1977).
53. S. W. Hawking, I. G. Moss, and J. M. Stewart, "Bubble collisions in the very early universe," *Phys. Rev.* **D26**, 2681–2693 (1982).
54. A. H. Guth and E. J. Weinberg, "Could the universe have recovered from a slow first order phase transition?" *Nucl. Phys.* **B212**, 321–364 (1983).
55. See, for example, E. W. Kolb and M. S. Turner, "Grand unified theories and the origin of the baryon asymmetry," *Annu. Rev. Nucl. Part. Sci.* **33**, 645–696 (1983).
56. V. A. Kuzmin, V. A. Rubakov, and M. E. Shaposhnikov, "On the anomalous electroweak baryon number nonconservation in the early universe," *Phys. Lett.* **155B**, 36 (1985).
57. L. McLerran, M. Shaposhnikov, N. Turok, and M. Voloshin, "Why the baryon asymmetry of the universe is $\sim 10^{-10}$," *Phys. Lett.* **256B**, 451 (1991).
58. V. A. Rubakov, "Electroweak baryon number violation at high energies," *Physica Scripta* **T36**, 194–198 (1991). Also published in *The Birth and Early Evolution of Our Universe*, eds. J. S. Nilsson, B. Gustafsson, and B.-S. Skagerstam (World Scientific, Singapore, 1991), pp. 194–198.
59. A. E. Nelson, D. B. Kaplan, and A. G. Cohen, "Why there is something rather than nothing: matter from weak interactions," *Nucl. Phys.* **B373**, 453–478 (1992).
60. P. J. E. Peebles, "The mean mass density of the universe," *Nature* **321**, 27–32 (1986).
61. N. Kaiser, G. Efstathiou, R. Ellis, C. Frenk, A. Lawrence, M. Rowan-Robinson, and W. Saunders, "The large-scale distribution of IRAS galaxies and the predicted peculiar velocity field," *Mon. Not. R. Astron. Soc.* **252**, 1–12 (1991).
62. A. Yahil, "The quest for Ω: comparison of density and peculiar velocity fields," in *The Early Universe and Cosmic Structures* (Proceedings of the XXV Rencontre de Moriond, March 4–11, 1990), eds. J. M. Alimi *et al.* (Editions Frontières, Gif-sur-Yvette, 1991), pp. 483–500.
63. E. R. Harrison, "Fluctuations at the threshold of classical cosmology," *Phys. Rev.* **D1**, 2726–2730 (1970).
64. Ya. B. Zel'dovich, "A hypothesis, unifying the structure and the entropy of the universe," *Mon. Not. R. Astron. Soc.* **160**, 1P–3P (1972).
65. E. L. Wright *et al.*, "Interpretation of the CMB anisotropy detected by the COBE differential microwave radiometer," *Astrophys. J. Lett.* **396**, L13–L18 (1992).
66. A. Vilenkin, "The birth of inflationary universes," *Phys. Rev.* **D27**, 2848 (1983).
67. A. A. Starobinskii, "Stochastic de Sitter (inflationary) stage in the early universe," in *Field Theory, Quantum Gravity and Strings*, eds. H. J. de Vega and N. Sánchez (Springer-Verlag Lecture Notes in Physics, Vol. 246, Berlin, 1986), pp. 107–126.
68. A. D. Linde, "Eternal chaotic inflation," *Mod. Phys. Lett.* **A1**, 81 (1986).
69. A. S. Goncharov, A. D. Linde, and V. F. Mukhanov, "The global structure of the inflationary universe," *Int. J. Mod. Phys.* **A2**, 561–591 (1987).

70. M. Aryal and A. Vilenkin, "The fractal dimension of inflationary universe," *Phys. Lett.* **199B**, 351–357 (1987).
71. A. H. Guth and S.-Y. Pi, "Quantum mechanics of the scalar field in the new inflationary universe," *Phys. Rev.* **D32**, 1899–1920 (1985).
72. H. Alfvén and O. Klein, "Matter and antimatter annihilation in cosmology," *Ark. Fys.* **23**, 187–194 (1962).

DO THE LAWS OF PHYSICS ALLOW US TO CREATE A NEW UNIVERSE?*

ALAN H. GUTH†

Center for Theoretical Physics
Laboratory for Nuclear Science and Department of Physics
Massachusetts Institute of Technology
Cambridge, Massachusetts 02139, USA

Essentially all modern particle theories suggest the possible existence of a false vacuum state — a metastable state with an energy density that cannot be lowered except by means of a very slow phase transition. Inflationary cosmology makes use of such a state to drive the expansion of the big bang, allowing the entire observed universe to evolve from a very small initial mass. A sphere of false vacuum in our present universe, if larger than a certain critical mass, could inflate to form a new universe which would rapidly detach from its parent. A false vacuum bubble of this size, however, cannot be produced classically unless an initial singularity is present from the outset. E. Farhi, J. Guven, and I have therefore explored the possibility that a bubble of subcritical size, which classically would evolve to a maximum size and collapse, might instead tunnel through a barrier to produce a new universe. We estimated the tunneling rate using semiclassical quantum gravity, and discovered some interesting ambiguities in the formalism.

1. The Cosmic Cookbook

During the past decade, a radically new picture of cosmology has emerged. The novelty of this picture is particularly striking when one considers the question of what it takes to produce a universe. According to the standard big bang picture of a decade ago, the visible universe could be assembled, at

*Based on a lecture given at Nobel Symposium 79, Gräftåvallen, Sweden, June 11–16, 1990, in *The Birth and Early Evolution of Our Universe*, eds. J. S. Nilsson, B. Gustafsson, and B.-S. Skagerstam (*Physica Scripta*, Vol. T36, and World Scientific, Singapore, 1991), pp. 237–246.

†This work was supported in part by funds provided by the US Department of Energy (DOE) under contract #DE-AC02-76ER03069.

$t \approx 1$ sec for example, by mixing approximately 10^{89} photons, 10^{89} e^+e^- pairs, 10^{89} $\nu\bar{\nu}$ pairs, 10^{79} protons, 10^{79} neutrons, and 10^{79} unpaired electrons. The total mass/energy of these ingredients is about 10^{65} grams $\approx 10^{32}$ solar masses $\approx 10^{10}\times$ present mass of visible universe. The mass is much larger than the present mass of the visible universe, because most of the energy is lost to gravitational potential energy as the universe expands.

With the advent of grand unified theories (GUT's) and inflationary cosmology, however, a much simpler recipe for a universe can now be formulated. (For a review of inflation, see any of Refs. 1–8.) To produce a universe at $t \approx 10^{-35}$ sec, the *only* necessary ingredient is a region of false vacuum. And the region need not be very large. For a typical GUT energy scale of $\sim 10^{14}$ GeV (which I will use for all the numerical examples in this paper), the minimum diameter is about 10^{-24} cm. The total mass/energy of this ingredient is only about 10 kg $\approx$ 10^{-29} solar masses $\approx$ $10^{-51}\times$ present mass of visible universe. This recipe sounds so easy that one cannot resist asking whether the laws of physics as we know them in principle permit the creation of a new universe by man-made processes. We believe that the answer to this question is yes,[9–11] but we do not yet have a definite answer. The problem turns out to be intimately embroiled in the ambiguities of quantum gravity. In this paper I will try to summarize our current understanding.

In considering this question, one difficulty is immediately obvious: the mass density of the required false vacuum is about 10^{75} g/cm^3. This mass density is certainly far beyond anything that is technologically possible, either now or in the foreseeable future. Nonetheless, for the purposes of this discussion I will whimsically assume that some civilization in the distant future will be capable of manipulating these kinds of mass densities. There are then some very interesting questions of principle that must be addressed in order to decide if the creation of a new universe is possible. While I will discuss these questions in terms of the possibility of man-made creation, I want to emphasize that the same questions will no doubt also have relevance to various natural scenarios that one could imagine. The question of man-made creation helps to simplify the discussion by focusing attention on the fundamental issue: what are the minimal conditions necessary for a new universe to be created?

The outline of this paper will be as follows. Section 2 discusses the classical evolution of a false vacuum bubble, and Sec. 3 will discuss the role of initial singularities in the classical evolution. Section 4 will be the main section of the paper, describing the formulation of a quantum gravity tunneling mechanism for producing a new universe. Section 5 will focus on a subtlety in the calculation — the evaluation of the subtraction term. Numerical results will be presented in Sec. 6, and the conclusions will be summarized and discussed in Sec. 7.

2. Evolution of a False Vacuum Bubble

A false vacuum is a peculiar state of matter that arises when a particle theory contains a scalar field ϕ with a potential energy function that has a local minimum which is different from the global minimum. A false vacuum is then a region of space in which $\phi = \phi_f$, where ϕ_f denotes the value of ϕ at the local minimum. The energy density of the false vacuum is then fixed at the value of the potential at the local minimum, which we denote by ρ_f. (I use units with $\hbar = c = 1$, so ρ_f is both an energy density and a mass density.) The energy–momentum tensor for this state is given by the Lorentz-invariant expression

$$T_{\mu\nu} = -\rho_f g_{\mu\nu} \,. \tag{1}$$

The energy density and pressure are then given by

$$\rho = \rho_f \tag{2}$$

and

$$p = -\rho_f \,. \tag{3}$$

By a false vacuum bubble, I mean a region of false vacuum surrounded by anything else. I will discuss in detail, however, only the simplest case: a spherical region of false vacuum surrounded by true vacuum, assumed to have zero energy density. In considering the evolution of such a region one is led immediately to a paradox. If the region of false vacuum is large enough, one expects that it would undergo inflation. An observer in the outside true vacuum region, on the other hand, would see the false vacuum region as a region of negative pressure. The pressure gradient would point inward, and the observer would not expect to see the region increase in size. The resolution of this paradox hinges on the dramatic distortion of spacetime that is caused by the false vacuum bubble.

Here I will outline the solution to the classical evolution problem, but a reader interested in the technical details will have to consult the literature.[12–18] The description given here follows the work that I did with Blau and Guendelman.[17]

The first step in solving the problem is to dissect it, dividing the spacetime into three regions. The exterior region is spherically symmetric empty space, for which the unique solution, in general relativity, is the well-known

Schwarzschild metric. The interior region consists of spherically symmetric false vacuum, and is required to be regular at $r = 0$. This spacetime also has a unique solution: de Sitter space. At the interface between these two regions is a domain wall — a region in which a scalar field is undergoing a transition between its true and false vacuum values. The solution that I will describe uses a thin-wall approximation, in which the thickness of the wall is assumed to be negligibly small compared to any other distance in the problem. In this approximation it can be shown that the surface energy density is equal to the surface tension and is independent of time, and we take this surface energy density σ as an additional parameter of the problem. The wall can then be described mathematically by a set of junction conditions[19] which are obtained by applying Einstein's equations to an energy–momentum tensor restricted to a thin sheet. (These equations are just the gravitational analogue of the well-known statement that the normal component of an electric field has a discontinuity of $4\pi\sigma$ at a sheet of surface charge density σ.) The evolution is completely determined by using these junction conditions to join the exterior forms of the metric.

The evolution of the bubble wall can be described by the function $r(\tau)$, where r is the radius of the bubble wall (defined as $\frac{1}{2\pi}$ times the circumference), and τ is the proper time as would be measured by a clock that follows the bubble wall. The junction conditions described above imply an equation of motion for $r(\tau)$ that can be cast into a form identical to that of a nonrelativistic particle moving in a potential. Using the definitions

$$\chi^2 \equiv \frac{8\pi}{3} G\rho_f \,, \tag{4}$$

$$\kappa \equiv 4\pi G\sigma \,, \tag{5}$$

and

$$\gamma \equiv \frac{2\kappa}{\sqrt{\chi^2 + \kappa^2}} \,, \tag{6}$$

the equations of motion can be written[17]

$$\left(\frac{dz}{d\tau'}\right)^2 + V(z) = E \,, \tag{7}$$

where z and τ' are rescaled radius and time variables defined by

$$z^3 \equiv \frac{\chi^2 + \kappa^2}{2GM} r^3 \tag{8}$$

and

$$\tau' \equiv \frac{\chi^2 + \kappa^2}{2\kappa}\,\tau \,. \tag{9}$$

E is related to the mass M of the physical bubble by

$$E \equiv \frac{-4\kappa^2}{(2GM)^{2/3}(\chi^2 + \kappa^2)^{4/3}}\,, \tag{10}$$

and the potential energy function $V(z)$ is given by

$$V(z) = -\left[\frac{1 - z^3}{z^2}\right]^2 - \frac{\gamma^2}{z}\,. \tag{11}$$

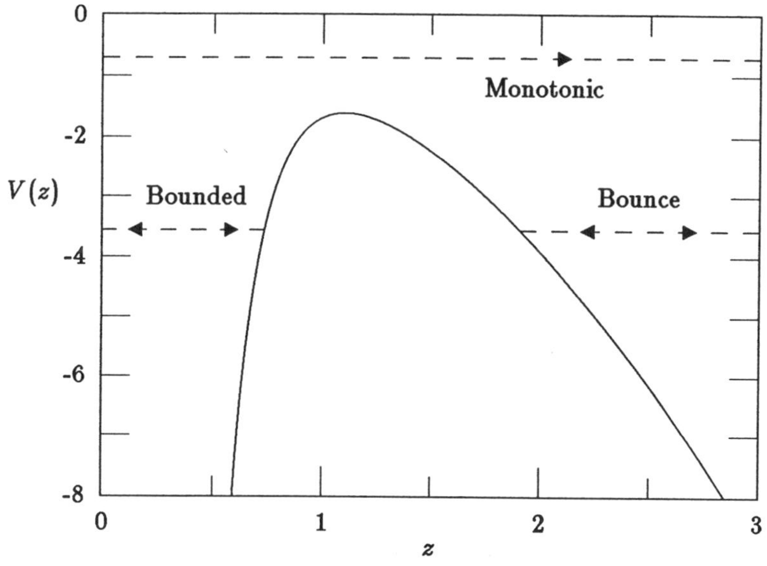

Fig. 1. In the thin-wall approximation the trajectory of the bubble wall is equivalent to the motion of a nonrelativistic particle in the potential energy curve V shown above. The energy of the fictitious particle is related to the mass of the false vacuum bubble; the energy increases with the mass, and approaches the top line of the diagram as $M \to \infty$. The curve was plotted with $\gamma = 1.3$.

A graph of $V(z)$ is shown as Fig. 1. As can be seen in this figure, there are three kinds of solutions. First, there are "bounded" solutions in which the bubble grows from $r = 0$ to a maximum size and then collapses. Second, there are "bounce" solutions. Here the bubble starts at infinite size in the asymptotic past, contracts to a minimum size, and then expands without limit. Finally, there are "monotonic" solutions — bubbles that start at zero size and grow monotonically. The monotonic solutions require a minimum mass, so that the energy of the fictitious particle is high enough to get over the potential barrier in Fig. 1. For small values of the surface energy density ($\sigma \ll \sqrt{\rho_f/G}$), this critical mass is given simply by

$$M_{cr} = \frac{4\pi}{3}\rho_f \chi^{-3} . \tag{12}$$

This expression can be interpreted as the mass density of the false vacuum multiplied by the volume of a Euclidean sphere with a radius equal to the de Sitter horizon distance, χ^{-1}. For typical GUT parameters, $M_{cr} \approx 10$ kg.

Having described the evolution of the bubble wall, I must still describe how the bubble wall is embedded in spacetime. For now I will describe only the behavior of the monotonic solutions. Since the exterior of the bubble is Schwarzschild space, it can be described by the familiar Schwarzschild coordinate system, with metric

$$ds^2 = -A_S dT_S^2 + A_S^{-1} dR^2 + R^2 d\Omega^2 , \tag{13}$$

where

$$A_S = 1 - \frac{2GM}{R} , \tag{14}$$

$d\Omega^2 = d\theta^2 + \sin^2\theta d\phi^2$, and M is the total energy of the bubble. Since these coordinates do not cover all of Schwarzschild space and since they have a coordinate singularity at $R = 2GM$, it is useful to introduce Kruskal–Szekeres[20] coordinates. The new coordinates (V_S, U_S) replace (T, R), while θ and ϕ are kept. These coordinates cover the entire manifold and have no coordinate singularities, and they also have the convenient property that lightlike lines lie at 45° to the vertical. The Schwarzschild R-coordinate is related to V_S and U_S by

$$\left(\frac{R}{2GM} - 1\right) e^{R/2GM} = U_S^2 - V_S^2 . \tag{15}$$

Lines of constant T_S are straight lines through the origin of the (V_S, U_S) plane, with

$$T_S = \begin{cases} 4GM \tanh^{-1}(V_S/U_S) & \text{if } |V_S/U_S| < 1 \\ 4GM \tanh^{-1}(U_S/V_S) & \text{if } |V_S/U_S| > 1. \end{cases} \tag{16}$$

The relation between the two coordinate systems is described more fully in Ref. 17. A diagram showing the relation between the Schwarzschild and the

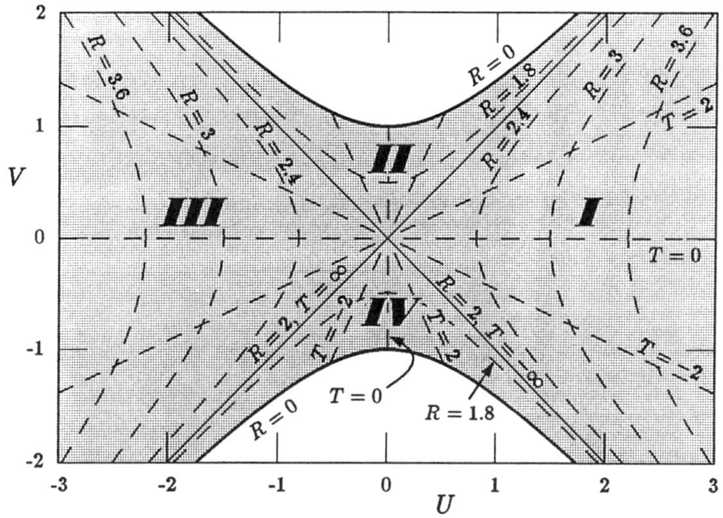

Fig. 2. A diagram of the Kruskal–Szekeres coordinate system. The four quadrants are labeled I, II, III, and IV, while the unshaded region is not part of the spacetime. Lightlike lines are at 45° to the vertical, and the future direction is shown as upward. The relation to the Schwarzschild coordinates is illustrated by lines of constant T and R, labeled in units of GM.

Kruskal–Szekeres coordinate systems is shown as Fig. 2. The metric for the Kruskal–Szekeres coordinates can then be written as

$$ds^2 = \frac{32(GM)^3}{R} e^{-R/2GM}(-dV_S^2 + dU_S^2) + R^2 d\Omega^2 , \tag{17}$$

where R is expressed implicitly as a function of V_S and U_S by using Eq. (15).

A spacetime diagram for the monotonic trajectory is shown as Fig. 3. The true vacuum region, to the right of the bubble wall, is shown in the standard Kruskal–Szekeres coordinates. The false vacuum region, to the left of the bubble wall, is shown in peculiar coordinates designed solely to allow the two halves of the diagram to fit together in the plane. The diagram is constructed so that lightlike lines lie at 45° to the vertical, but the metric is highly distorted. In particular the exponential expansion of the false vacuum region, which occurs as one moves upward and to the left in the diagram, is completely hidden by the distortion of the metric.

The physical meaning of a spacetime diagram of this type can be seen most clearly by examining a sequence of equal-time slices. Figure 4 shows

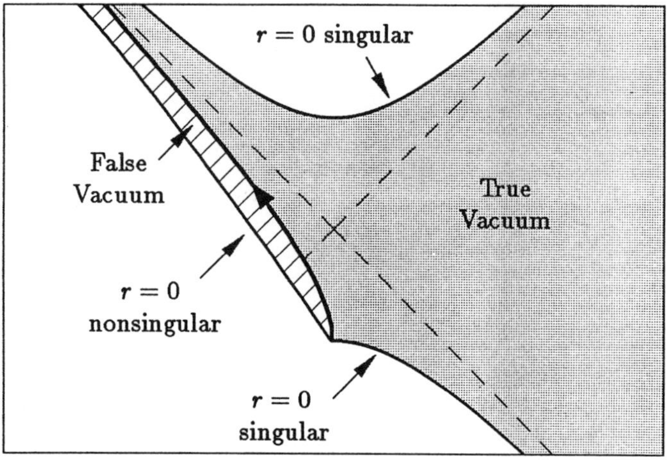

Fig. 3. A spacetime diagram of a monotonic false vacuum bubble solution. Angular coordinates are suppressed, and the diagram is plotted so that lightlike lines are at 45°. The bubble wall is shown as a heavy line with an arrow on it. The true vacuum region (shaded) is to the right of the bubble wall, and the false vacuum region (diagonal lines) is to the left. The diagram shows the initial (lower) and final (upper) $r = 0$ singularities, and also a nonsingular $r = 0$ line (i.e., the center of a spherical coordinate system) that runs along the left edge.

the positions of four slices, labeled (a), (b), (c), and (d), and Fig. 5 shows a representation of each slice. For purposes of illustration, Fig. 5 shows only two of the three spatial dimensions. Since the spaces of interest are spherically symmetric, this results in no loss of information. The two-dimensional sheet is shown embedded in a fictitious third dimension, so that the curvature can be visualized. Figure 5(a) shows a space which is flat at large distances, but which has a singularity at the origin. In Fig. 5(b) a small, expanding region of false vacuum has appeared at the center, replacing the singularity. The false vacuum region is separated from the rest of space by a domain wall. Figure 5(c) shows the false vacuum region beginning to swell. Note, however, that the swelling takes place by the production of new space; the plane of the original space is unaffected. The false vacuum region continues to inflate, and it soon disconnects completely from the original space, as shown in Fig. 5(d). As was shown by Sato *et al.*[12] in 1981, it forms a new, isolated closed universe.

Note, by the way, that Fig. 5 shows clearly how the paradox raised at the beginning of this section is resolved. The net force on the bubble wall points

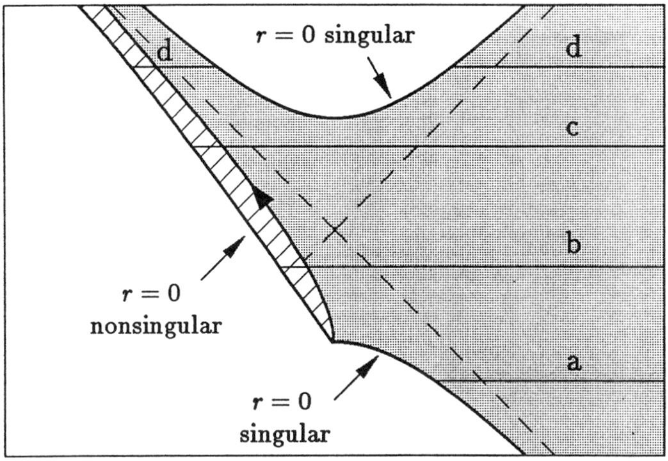

Fig. 4. Horizontal lines indicating spacelike hypersurfaces to be illustrated in Fig. 5.

from the true vacuum region to the false vacuum region, as expected. Due to the inversion shown in Figs. 5(c) and 5(d), however, this force causes the bubble wall to expand, rather than contract.

To summarize, the false vacuum bubble appears from the outside to be a black hole. From the inside, however, it appears to be an inflating region of false vacuum, with new space being created as the region expands. The region completely disconnects from the original spacetime, forming a new, isolated closed universe.

Although the problem that has been solved is very idealized, it nonetheless appears to contain the essential physics of more complicated inhomogeneous spacetimes. The paradox discussed at the beginning of this section will exist whenever an inflating region is surrounded by noninflating regions, and the qualitative behavior of the system seems to be determined by the way in which this paradox is resolved. Thus, one concludes that if inflation occurred in an inhomogeneous universe, then many isolated closed universes would have been ejected.

Furthermore, even if inflation somehow began in a completely homogeneous way, one still expects the universe to break apart into a host of disconnected universes. The reason stems from the intrinsic nonuniformity, on very large scales, of the decay of the false vacuum.[21–23] The decay of the false vacuum

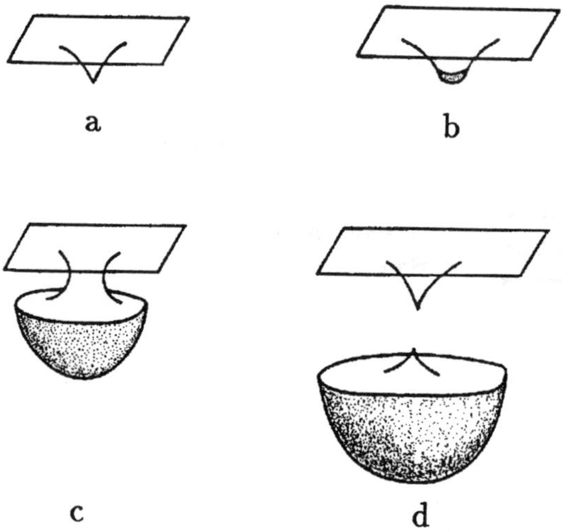

Fig. 5. The evolution of a monotonic false vacuum bubble solution. Each lettered diagram illustrates a spacelike hypersurface indicated in Fig. 4. The diagrams are drawn by suppressing one dimension of the hypersurface and embedding the resulting two-dimensional surface in a fictitious three-dimensional space so that the curvature can be displayed. The false vacuum region is shown as shaded. Note that diagram (d) shows a new universe detaching from the original spacetime.

occurs exponentially,[a] like most other decay processes, but for inflation to be successful the parameters must be arranged so that the exponential decay constant is slow compared to the exponential expansion rate. This implies that the total volume of false vacuum *increases* with time. Thus, no matter how long one waits there will still be regions of false vacuum. These regions have no reason to be spherical, but the arguments of the previous paragraph lead one to expect a high likelihood of producing disconnected universes.

3. Initial Singularities as an Obstacle to Universe Creation

Figures 4–5 illustrate the creation of a new universe, but there is one undesirable feature. The sequence begins with an initial singularity, shown as the lower $r = 0$ singularity in Fig. 4, and as part (a) of Fig. 5. Although an initial singularity is often hypothesized to have been present at the big bang, there do not appear to be any initial singularities available today. So we ask

[a]I have studied a simplified but exactly soluble model of a slow-rollover phase transition with S.-Y. Pi in Ref. 24.

whether it is possible to intervene in some way, to modify the early stages of this picture, so that an inflationary universe could be produced *without* an initial singularity. This question has been addressed at both the classical and quantum levels.

At the classical level Farhi and I[25] have shown that, under reasonable assumptions, the initial singularity cannot be avoided. Any false vacuum bubble which grows to become a universe necessarily begins from an initial singularity.

The argument rests on an application of the Penrose theorem.[b] The inflationary solutions are very rapidly expanding, and the Penrose theorem implies that such rapid expansion can result only from an initial singularity. (The Penrose theorem is more widely known in a form that is the time-reverse of the present application: if a system is collapsing fast enough, there is no way to avoid the collapse to a singularity.)

The application of the Penrose theorem involves two technical loopholes. First, if the final bubble is not spherically symmetric, then we have not been able to show that the Penrose theorem applies. We believe that this shortcoming, however, is probably the result of our own limitations, and does not provide a way to avoid the theorem. Second, if a material can be found for which the magnitude of the pressure exceeds the energy density, then the Penrose theorem would not apply. In quantum field theories it is possible to construct states that have this property, but it is not clear if a large enough spacetime region of this type can be attained.

4. The Creation of a Universe by Quantum Tunneling

Although the Penrose theorem implies that a new universe cannot be created classically without an initial singularity, the Penrose theorem does not apply in the realm of quantum mechanics. Since the theorem is derived from the classical equations of motion, one is always free to look for violations at the quantum level. Farhi, Guven, and I[9] have therefore turned our attention to whether quantum physics allows the creation of an inflationary universe without an initial singularity. In particular, we have explored the following recipe. Suppose a small bubble of false vacuum (with mass less than the critical mass M_{cr}) is created and caused to expand at a moderate rate. Since the bubble is not expanding rapidly, the Penrose theorem does not preclude its production by classical processes without an initial singularity. We have not explored in detail the mechanisms by which such a region might be created,

[b] J. Bardeen, W. Israel, W. Unruh, and R. Wald are to be thanked for pointing out to us the relevance of this theorem.

but presumably it could be created either by supercooling from high temperatures or perhaps by compressing a gas of fermions that couple to the scalar field. If such a bubble were allowed to evolve classically, it would correspond to one of the bounded solutions, as discussed in the context of Fig. 1. It would expand to a maximum size and then the pressure gradient would halt the expansion and cause the bubble to collapse. By quantum processes, however, one might imagine that the bubble could tunnel through the potential energy barrier shown in Fig. 1, becoming a bounce solution that would continue to grow until eventually the false vacuum decayed. The late-time behavior of this bounce solution would strongly resemble that shown in Figs. 5(c) and 5(d). Although no fully satisfactory theory of quantum gravity exists, we have attempted to *estimate* the tunneling amplitude by using a semiclassical (WKB) approximation.

Specifically, we used the same kind of Euclidean field theory technique that was used by Coleman and De Luccia[26] to calculate the decay rate of the false vacuum in curved spacetime. (See also Sec. 6 of Ref. 27, which includes a discussion of a spacetime region that was omitted in the original reference.) That is, we assume that the amplitude for a transition from one three-geometry to another is well-approximated by $e^{iI_{\rm cl}/\hbar}$, where $I_{\rm cl}$ is the action of the classical solution to the field equations which interpolates between the two three-geometries. If no real-time solution exists then we seek a Euclidean four-geometry that solves the imaginary time field equation and whose boundary is the two three-geometries of interest. The tunneling amplitude is then estimated as $e^{-I_E/\hbar}$, where I_E is the properly subtracted classical action of the Euclidean solution — that is, it is the action of the solution, minus the action of a configuration that remains static at the initial state of the tunneling process for the same Euclidean time as the solution requires for its transit. (The motivation for this subtraction term will be discussed in Sec. 5.)

Thus, the key step that remains is to construct the Euclidean solution that interpolates between the initial and final states of the tunneling.

In Sec. 2 we discussed the Kruskal–Szekeres coordinate system for Schwarzschild space. To continue, we must similarly define our coordinate system for the de Sitter part of the problem. One well-known system is the static coordinates, closely analogous to the Schwarzschild form of the metric in Schwarzschild space. The static metric is given by

$$ds^2 = -A_D dT_D^2 + A_D^{-1} dR^2 + R^2 d\Omega^2 \,, \tag{18}$$

where

$$A_D = 1 - \chi^2 R^2 \,. \tag{19}$$

These coordinates have a coordinate singularity at $R = 1/\chi$, and they fail to cover the entire de Sitter manifold. Both of these shortcomings can be avoided by replacing T_D and R by the Gibbons–Hawking[28] coordinates (V_D, U_D), which are analogous to the Kruskal–Szekeres coordinate system. Once again θ and ϕ are kept. The static coordinate R is related to V_D and U_D by

$$\frac{1-\chi R}{1+\chi R} = U_D^2 - V_D^2 \,. \tag{20}$$

Lines of constant T_D are again straight lines through the origin, with

$$T_D = \begin{cases} \chi^{-1}\tanh^{-1}(V_D/U_D) & \text{if } |V_D/U_D| < 1 \\ \chi^{-1}\tanh^{-1}(U_D/V_D) & \text{if } |V_D/U_D| > 1 \,. \end{cases} \tag{21}$$

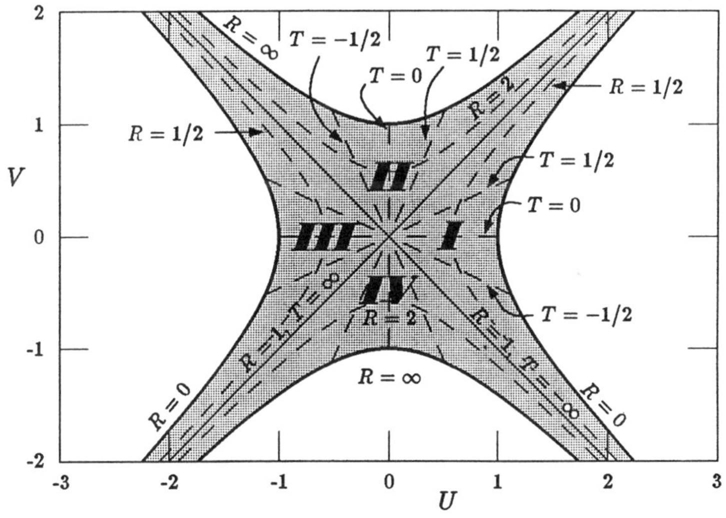

Fig. 6. A diagram of the Gibbons–Hawking coordinate system of de Sitter space, which is a close analogue of the Kruskal–Szekeres coordinate system of Schwarzschild space. The four quadrants are labeled I, II, III, and IV, and again the spacetime does not exist outside the shaded region. The relation to the static de Sitter coordinates is shown by lines of constant R and T, labeled in units of χ^{-1}.

The relation between the two de Sitter coordinate systems is illustrated in Fig. 6, and it is also described more fully in Ref. 17. The metric in the Gibbons–Hawking system is given by

$$ds^2 = \chi^{-2}(1+\chi R)^2(-dV_D^2 + dU_D^2) + R^2 d\Omega^2 \,, \tag{22}$$

where R is determined by Eq. (20). Again lightlike lines travel at 45°. The presence of the term $R^2 d\Omega^2$ in all of the metrics that we have introduced implies that R always has the same meaning: the rotational symmetry relates any given point to a sphere of points, and R is equal to $\frac{1}{2\pi}$ times the circumference of this sphere.

We are now ready to describe the classical trajectories involved in the tunneling process. Specifically, we wish to estimate a tunneling rate from a bounded trajectory to a bounce trajectory, as shown in Fig. 1. The form of the trajectories depends on the mass of the bubble, and we begin by considering a bubble just below the critical mass. A trajectory of this type is shown in the upper part of Fig. 7. The left and right halves of the diagram show the trajectory of the bubble wall as seen in the interior and exterior coordinate systems, respectively.

The initial slice for the tunneling problem is taken to be the $T_S = 0$ surface in the Schwarzschild region and the $T_D = 0$ surface in the de Sitter region, as shown in the diagram. The final trajectory is shown in the lower part of Fig. 7. The final slice is again taken along the $T_S = 0$ and $T_D = 0$ surfaces. Note that the extrinsic curvature vanishes on both the initial and final slices. This fact is the geometric generalization of the condition that $\dot{q} = 0$ at the turning points marking the beginning and end of a tunneling trajectory in a one-degree-of-freedom quantum mechanics problem.

To construct the Euclidean interpolation, we begin by considering the de Sitter half of the problem. Euclidean de Sitter space is a four-sphere with radius $1/\chi$. The Euclidean continuation of the Gibbons–Hawking form of the metric (22) is

$$ds_E^2 = \chi^{-2}(1 + \chi R)^2 (dV_{DE}^2 + dU_{DE}^2) + R^2 d\Omega^2 \,, \tag{23}$$

where $0 \leq R \leq 1/\chi$ and

$$\frac{1 - \chi R}{1 + \chi R} = U_{DE}^2 + V_{DE}^2 \,. \tag{24}$$

The static time variable is related to the angle in the U_{DE}–V_{DE} plane, with

$$T_{DE} = \chi^{-1} \tan^{-1}(V_{DE}/U_{DE}) \,. \tag{25}$$

The entire space is represented by a disk $U_{DE}^2 + V_{DE}^2 \leq 1$ in the U_{DE}–V_{DE} plane. Each point in the interior of the disk represents a two-sphere of radius R, where R varies from 0 at the boundary to $1/\chi$ at the origin.

To construct the Euclidean interpolation, first note that the hypersurfaces i and f from Fig. 7 can be copied without modification onto the Euclidean diagrams. Crudely speaking, this can be seen by noting that the vanishing of

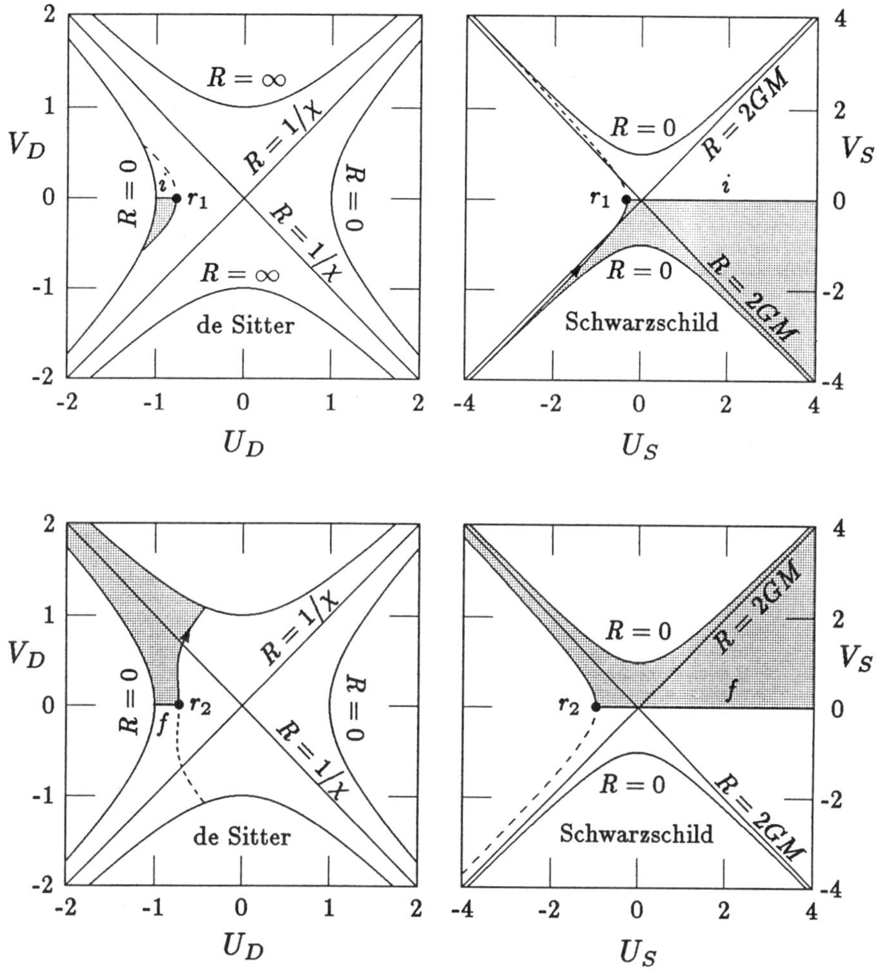

Fig. 7. Trajectories for the tunneling problem, for a bubble mass just below M_{cr}. The upper diagrams show the initial trajectory. Full de Sitter and Schwarzschild spaces are shown, with shaded regions indicating the physical spacetime. The initial slice for the tunneling problem is shown as a horizontal line labeled i. The lower diagrams indicate the final trajectory, with the final slice shown as a horizontal line labeled f. (This figure and the following two are drawn for $\gamma = 1.95$ and $M/M_{cr} = .97$).

the extrinsic curvature means that there are no derivatives of the metric in the time direction. Since the time derivative vanishes, it does not matter whether the surfaces evolve in Euclidean or Lorentzian time. It is then straight-

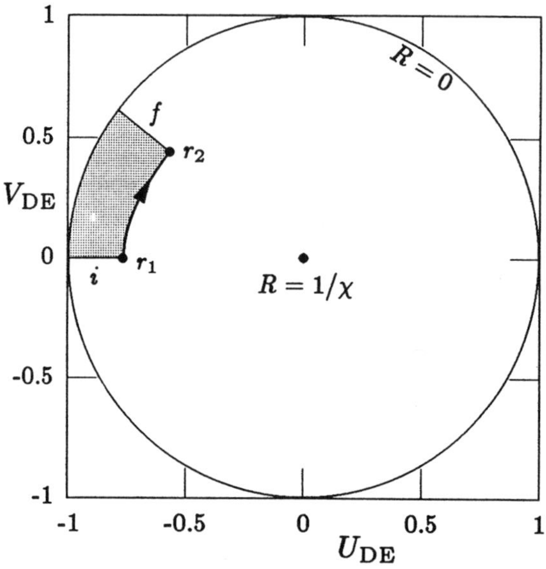

Fig. 8. The de Sitter part of the interpolating Euclidean four-geometry. The shaded area shows the region swept out as the initial hypersurface i evolves to become the final hypersurface f. The bubble wall trajectory is shown as a heavy line with an arrow on it.

forward to calculate the Euclidean trajectory of the domain wall, by analytically continuing the normal equations of motion.

The parts of the initial and final surfaces within the de Sitter part of the space, and the interpolating trajectory of the bubble wall, are shown in Fig. 8. From this figure we can see that the de Sitter half of the diagram has a simple interpretation. We can visualize the initial surface evolving into the final surface, while at each stage of the evolution the surface extends from the bubble wall to the edge of the de Sitter disk, where $R = 0$. The region swept out by this motion, shown as a shaded region in the diagram, is the de Sitter part of the interpolating four-geometry. This part of the problem is straightforward and trouble-free.

The Schwarzschild part of the problem, however, does not work out nearly so cleanly. We begin by Euclideanizing the metric itself, which is straightforward. In Kruskal–Szekeres coordinates, the Euclideanized version of the metric (17) becomes

$$ds_E^2 = \frac{32(GM)^3}{R}\, e^{-R/2GM}(dV_{SE}^2 + dU_{SE}^2) + R^2 d\Omega^2 \,, \tag{26}$$

where

$$\left(\frac{R}{2GM} - 1\right)e^{R/2GM} = U_{SE}^2 + V_{SE}^2 \,. \tag{27}$$

The Schwarzschild time variable is related to the angle in the U_{SE}–V_{SE} plane by

$$T_{SE} = 4GM \tan^{-1}(V_{SE}/U_{SE}) \,. \tag{28}$$

Note that $R = 2GM$ at the origin of the U_{SE}–V_{SE} plane, and grows monotonically as one moves outward. Thus, a diagram of Euclidean Schwarzschild space (with θ and ϕ suppressed) consists of a central dot at which $R = 2GM$, and concentric circles of constant R surrounding it.

Again it is straightforward to copy the initial and final hypersurfaces onto the Euclidean diagram, and to find by analytic continuation the bubble wall trajectory that interpolates between them. The results of these calculations are illustrated in Fig. 9. Since the bubble wall trajectory subtends an angle greater than π, it crosses both the initial and final surfaces. In addition, the initial and final surfaces cross each other. We have carried out rather extensive numerical calculations showing that the subtended angle always has a magnitude greater than π, so the qualitative features of the diagram are not particular to the chosen values of the parameters.

As in the de Sitter case, we can visualize the initial surface evolving into the final surface; this time, however, it is more complicated. We might imagine, for example, that the surface pivots about the origin, with one end extending to the bubble wall, and the other end to infinity. However, since some points will be crossed more than once by this motion, we cannot identify a region swept out. There is no simple analogue to the shaded region shown for the de Sitter case in Fig. 8.

The difficulty in interpreting the solution shown in Fig. 9 becomes even more striking if one imagines moving slightly away from the thin-wall limit. In that case one expects to be able to define a continuously varying scalar field on the Euclidean manifold. One does not know, however, how to treat the points where the bubble wall trajectory intersects the initial or final surfaces. If one thinks of these points as lying on the initial or final surfaces, then they are well into the Schwarzschild region and the scalar field should have its true vacuum value. On the other hand, if the points are considered to be on the domain wall, then the scalar field in the vicinity should be somewhere between the true vacuum and false vacuum values.

It is possible, nonetheless, to define a covering number for each point, as is described in the caption to Fig. 9. As can be seen on the diagram, the covering number takes on the values 1, 0, and -1 for this situation.

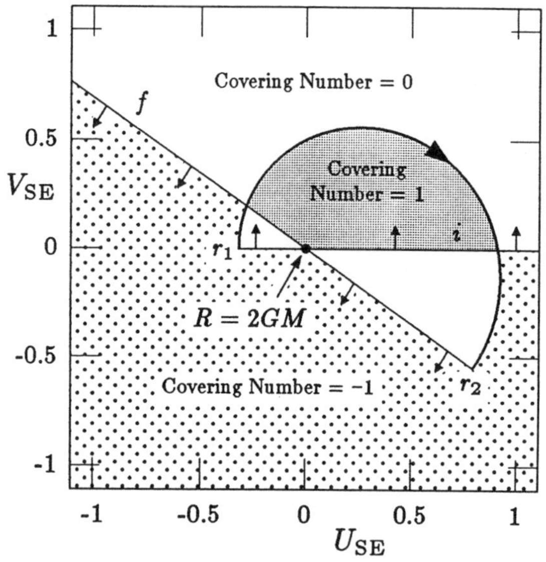

Fig. 9. The Schwarzschild part of the interpolating Euclidean four-geometry. The bubble wall trajectory, shown as a heavy line with an arrow on it, crosses both the initial and final surfaces, which also cross each other. Consequently, one cannot identify the region swept out as the initial hypersurface i evolves to become the final hypersurface f. Instead one must define a covering number, an integer that indicates the net number of times a point is covered by the evolving hypersurface. The small arrows on the initial hypersurface i and the final hypersurface f indicate the future direction, defined by the arrow of evolution of the bubble wall trajectory at the point where the hypersurface meets the trajectory. To define the covering number, imagine the evolution of the initial hypersurface into the final. The covering number for any point is the number of times the point is crossed by the evolving surface with the future side first, minus the number of times it is crossed with the past side first. The region with covering number 1 is shown shaded, the region with covering number -1 is shown dotted, and the region with covering number 0 is shown white.

The fact that the covering numbers take values other than 0 and 1 implies that we have not found a genuine four-dimensional Euclidean manifold that interpolates between the initial and final three-geometries. We are not sure what this means. Nonetheless, we have at least constructed a sequence of three-geometries that interpolate between the initial and final three-geometries, and perhaps this is enough to give a valid semiclassical approximation to the tunneling amplitude. In our paper[9] we conjecture that this approach is correct, and proceed to calculate the tunneling amplitude. We use the word "pseudo-manifold" to describe a space with covering numbers defined on it, and we define an action that is weighted by the covering numbers. We can show that the action defined in this way, when varied, yields the correct Euclidean equations of motion.

The possibility of a multisheeted covering comes to mind, especially when thinking about the difficulty in defining a continuously varying scalar field. The possibility of a nonsingular multisheeted covering, however, is excluded by the fact that the underlying space is simply connected. To have a multisheeted structure, there must be a rule that allows us to determine, for any closed path, whether the path returns to the original sheet or a different one. If all closed paths can be continuously deformed into a trivial path, as is the case here, then continuity implies that all closed paths must return to the original sheet.

Although a nonsingular multisheeted covering is excluded, a new coordinate system giving a *singular* multisheeted covering can nonetheless be constructed. In fact, such a singular multisheeted covering is probably the most convenient way to represent the pseudomanifold. The argument in the previous paragraph is bypassed, since the coordinates of the covering will not be continuous functions of the coordinates of the original Euclidean Schwarzschild and de Sitter manifolds. Then, as a path is continuously deformed in the Schwarzschild or de Sitter manifold, the image of the path in the new coordinates can jump discontinuously from one sheet to another. Such a coordinate system can be constructed, for example, by first parametrizing the bubble wall trajectory by the proper time variable τ_E. Then for each point $P(\tau_E)$ along the trajectory we choose a surface that extends from $R = 0$ in the de Sitter region to the point P, and then in the Schwarzschild region from P to $R = \infty$. (There is no requirement, however, that R vary monotonically along this surface.) We further insist that the surfaces are chosen continuously in τ_E, and that the initial and final surfaces coincide with the initial and final surfaces of the tunneling problem. Assign the coordinate τ_E to each point on the surface chosen through $P(\tau_E)$. Now define on each surface a coordinate σ, extending from 0 at $R = 0$ to ∞ at $R = \infty$. We insist that σ varies continuously and monotonically along the surface, and that σ is chosen to be continuous in τ_E. We let $\sigma(\tau_E)$ indicate the value of the σ-coordinate at the point $P(\tau_E)$. Note that some points in the original Schwarzschild and de Sitter manifolds will be covered several times by this procedure, and other points will not be covered at all—if a sign is attributed to each covering of this evolving surface as before, then the net number of times a point is covered is just the covering number defined earlier. The coordinates $\xi^\mu \equiv (\tau_E, \sigma, \theta, \phi)$ define the multisheeted covering of the pseudomanifold.

We now assign a metric $\tilde{g}_{\mu\nu}$ to the ξ-space as follows:

$$\tilde{g}_{\mu\nu} = \begin{cases} \dfrac{\partial x^\lambda}{\partial \xi^\mu} \dfrac{\partial x^\rho}{\partial \xi^\nu} g_{\lambda\rho}^{DE}(x(\xi)) & \text{if } \sigma < \sigma(\tau_E) \\ \dfrac{\partial x^\lambda}{\partial \xi^\mu} \dfrac{\partial x^\rho}{\partial \xi^\nu} g_{\lambda\rho}^{SE}(x(\xi)) & \text{if } \sigma < \sigma(\tau_E)\,, \end{cases} \tag{29}$$

where $g_{\lambda\rho}^{DE}$ and $g_{\lambda\rho}^{SE}$ denote the Euclidean de Sitter and Schwarzschild metrics, respectively. Note that on each $\tau_E = constant$ surface the metric has only a single de Sitter region ($\sigma < \sigma(\tau_E)$) and a single Schwarzschild region ($\sigma > \sigma(\tau_E)$), even if the surface crosses the domain wall in Fig. 9 more than once. Note also that the junction conditions already imposed on the domain wall trajectory imply that the metric $\tilde{g}_{\mu\nu}$ describes a space that has no discontinuity at $\sigma = \sigma(\tau_E)$.

In the multisheeted coordinate system the action can be taken to have the standard form, except that the volume measure $\sqrt{\tilde{g}(\xi)}$ is defined by

$$\sqrt{\tilde{g}(\xi)} \equiv \det\left(\frac{\partial x}{\partial \xi}\right)\sqrt{g(x)} \tag{30}$$

and is allowed to have either sign, depending on the sign of $\det(\partial x/\partial\xi)$. The ξ-space is therefore not a Riemann manifold, since the determinant of the metric can vanish and change sign. It can be shown, however, that the variation of the action defined in this way leads to the correct Euclidean equations of motion. Furthermore, in this formalism one can define a scalar field on the space without encountering the ambiguities mentioned above.

5. Subtletlies in the Subtraction Term

One particularly subtle point in the tunneling problem is the calculation of the subtraction term, which will be explained in this section. If the reader wishes to skip this fine point, he/she can jump to Sec. 6 without any loss of continuity.

To see the origin of the subtraction term, one can consider a tunneling problem in one-dimensional nonrelativistic quantum mechanics. If a particle tunnels from one classical turning point A to another classical turning point B, the WKB amplitude can be calculated by first solving for the imaginary-time classical path $x(\tau)$, where τ is related to the time variable t by $\tau = it$. Defining a Euclidean momentum by $p_E \equiv -ip$, where p is the ordinary canonical momentum, the WKB exponent can be written as an integral over the classical path:

$$I = \int_A^B p_E dx \,. \tag{31}$$

Using the relationship between the Euclideanized Lagrangian L_E and the Euclideanized Hamiltonian H_E, this expression can be rewritten as

$$I = \int_{\tau_A}^{\tau_B} p_E \frac{dx}{d\tau} d\tau \tag{32a}$$

$$= \int_{\tau_A}^{\tau_B} d\tau \left[L_E \left(x, \frac{dx}{d\tau} \right) - H_E \right] \tag{32b}$$

$$= \int_{\tau_A}^{\tau_B} d\tau \, L_E \left(x, \frac{dx}{d\tau} \right) - \int_{\tau_A}^{\tau_B} d\tau \, L_E(x_A, 0) \,. \tag{32c}$$

In writing Eq. (32c), we made use of the fact that H_E has a constant value along the classical trajectory, and so it can be replaced by its value at the start of the trajectory. Note that the subtraction term, the second term in Eq. (32c), can be described as the action of a static configuration that stays fixed at the entry point x_A, and exists for the same length of time as the classical solution.

The subtlety in applying this principle to the false vacuum bubble tunneling problem is the fact that there is more than one relevant time variable. Should we use the de Sitter time, the Schwarzschild time, or maybe the proper time? The answer can be found by setting up a canonical formalism, so that the manipulations of Eq. (32) can be carried out. Such a canonical formulation can be obtained by starting with the well-known canonical formulation of the field theory, and then truncating to the space of the thin-wall configurations. To carry out this truncation, one first adopts a recipe for the slicing of spacetime, so that each segment of the bubble wall trajectory can be associated with a region of spacetime. The action for a segment of the trajectory is then taken to be the integral of the field theory Lagrangian, integrated over the associated spacetime region.

Note, however, that the classical variational principle requires that the fields be held constant on the boundary of the spacetime region. For this problem, the boundary includes the surface at spatial infinity. Since the Schwarzschild time interval can be determined by measurements at spatial infinity, the variational principle requires that the bubble wall trajectory be varied so that the Schwarzschild time interval is held fixed. This means that the de Sitter time interval and the proper time interval will not be fixed as the trajectory is varied, and therefore only the Schwarzschild time can be used as a canonical time variable. Thus, it is the Schwarzschild time interval that must be used to calculate the subtraction term.

6. Numerical Results

We have used our definition of the action to estimate the tunneling amplitude as a function of the various parameters in the problem, and we have found that it behaves very reasonably: the tunneling action decreases monotonically to zero as the bubble mass M approaches the critical mass $M_{\rm cr}$ at which tunneling would not be necessary, and it diverges monotonically as the gravitational constant $G \to 0$. The action is negative definite by the standard sign conventions,[c] which some physicists will regard as an indication of error. We, however, are not convinced that the sign should be a matter of concern. In any quantum mechanics tunneling problem, the same Schrödinger equation describes tunneling in both directions through the barrier. Thus the equation is solved by both exponentially decaying and exponentially growing wave functions, and one simply has to be sensible enough to choose the right solution for the problem of interest.

In a pair of recent papers, Fischler, Morgan, and Polchinski[10,11] have calculated an amplitude for this same process, using a Hamiltonian method that is very different from the method we used. In their formalism they find no inconsistencies, and their answer is identical to ours. We regard this agreement as a strong indication that the calculation is correct, or at least that it is the most sensible calculation that one can do in the absence of a true theory of quantum gravity.

The final result is obtained by a numerical integration, and it is shown graphically in Fig. 10. The (subtracted) Euclidean tunneling action I_E depends on $M/M_{\rm cr}$, and also on the dimensionless parameter γ, defined in Eq. (6). For typical GUT parameters, $\gamma \approx 10^{-4}$.

As a rough estimate, the action I_E is of order $1/(G\chi^2)$, as long as $G\sigma^2$ is smaller than or comparable to $\rho_{\rm f}$, and M is not too near $M_{\rm cr}$. For typical grand unified theory parameters, this would give an outrageously small tunneling probability, such as

$$10^{-(10^{18})} .$$

Even with this small probability, however, there might still be a large probability of an event of this sort occurring somewhere in a universe that has undergone a large amount of inflation. Thus, the possibility of a chain reac-

[c] We take as "standard" the convention that the kinetic terms of the matter fields in the Euclideanized Lagrangian are positive.

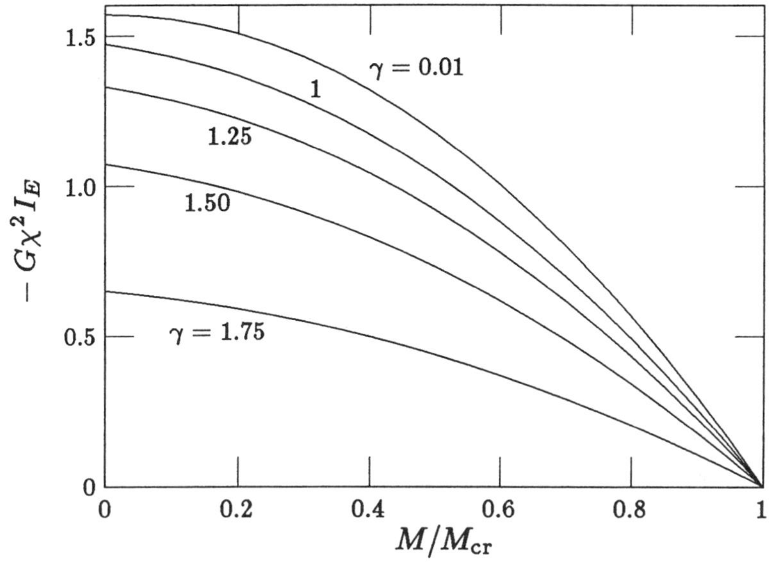

Fig. 10. Graph of $-G\chi^2 I_E$, where I_E is the Euclidean tunneling action. It is shown as a function of M/M_{cr}, for various values of the parameter γ.

tion by which one universe produces more than one universe is not obviously ruled out by this estimate. On the other hand, if we are talking about creating a universe in a hypothetical laboratory, then a probability this small must be considered equivalent to zero. Nonetheless, the numerical estimate of the tunneling probability is extraordinarily sensitive to the energy density ρ_f of the false vacuum. If there exists a false vacuum with an energy density near the Planck scale, which is certainly not excluded by anything we know, then the tunneling probability would be of order one.

7. Conclusions

In setting up the Euclidean formulation of the false vacuum bubble tunneling problem, we have found that no true Euclidean interpolating manifold exists. There is no difficulty or ambiguity in analytically continuing the bubble wall trajectory into the Euclidean regime, but when this trajectory is plotted on a Euclidean spacetime diagram it is found to cross both the initial and final surfaces of the tunneling problem. These intersection points prevent a conventional manifold interpretation.

We admit that we are not sure what the absence of a true interpolating manifold implies about the tunneling problem. Perhaps it indicates that the stationary phase method has failed, perhaps it indicates that one cannot extrapolate the thin-wall approximation into the Euclidean regime, or perhaps it is a suggestion that tunneling is for some reason forbidden.

We find it difficult to believe, however, that the tunneling process is forbidden, since there is no barrier to constructing a well-defined manifold (with either Lorentzian or Euclidean signature) that interpolates between the initial and final states. Such a manifold is not a solution, but it would constitute a path contributing to the functional integral. Furthermore, since any small variation about such a path would also contribute, the measure of these paths appears naively to be nonzero. The amplitude would then be nonzero unless the various paths conspire to cancel each other, as they do for an amplitude that violates a conservation principle associated with a symmetry. In the present case, however, there is no apparent symmetry or conservation law at work. We therefore conjecture that the tunneling process is allowed, and that the semiclassical approximation is valid. The alternative calculations of Fischler, Morgan, and Polschinski[10,11] add considerable weight to this point of view.

Although no Euclidean interpolating manifold exists, we find that it is nonetheless possible to generalize the notion of a manifold to describe a well-defined Euclidean interpolation. In our paper we defined an object that we called a "pseudomanifold," which we described in two alternative ways: by the use of covering numbers, and by the use of a singular multisheeted covering. In the second description the pseudomanifold closely resembles a true manifold, except that $\sqrt{g}$ is allowed to vanish and to change sign. The action of the pseudomanifold is defined by the usual expression for the Euclidean action, except that $\sqrt{g}$ is not positive definite.

If our semiclassical result is correct, then it appears to raise an important issue in quantum gravity: how does a pseudomanifold arise in a quantum gravity path integral? It might mean that such objects occur in the physical definition of the path integral, or it might mean that they arise as saddle points which are obtained by the distortion of integration contours in the complex plane.

In any case, this work seems to indicate that there are ambiguities in semiclassical Euclidean quantum gravity that are yet to be resolved.

References

1. A. D. Linde, *Rep. Prog. Phys.* **47**, 925 (1984).
2. A. D. Linde, in *300 Years of Gravitation*, eds. S. W. Hawking and W. Israel (Cambridge University Press, Cambridge, England, 1987), p. 604.
3. R. H. Brandenberger, *Rev. Mod. Phys.* **57**, 1 (1985).

4. M. S. Turner, in *Architecture of Fundamental Interactions at Short Distances*, eds. P. Ramond and R. Stora (North-Holland, Amsterdam, 1987), p. 513.
5. P. J. Steinhardt, in *High Energy Physics, 1985, Volume 2*, Proceedings of the Yale Theoretical Advanced Study Institute, eds. M. J. Bowick and F. Gürsey (World Scientific, Singapore, 1986), p. 567.
6. S. K. Blau and A. H. Guth, in *300 Years of Gravitation*, cited above, p. 524.
7. L. F. Abbott and S.-Y. Pi, eds., *Inflationary Cosmology* (World Scientific, Singapore, 1986).
8. A. D. Linde, *Particle Physics and Inflationary Cosmology* (Gordon and Breach, Boston, 1990).
9. E. Farhi, A. H. Guth, and J. Guven, *Nucl. Phys.* **B339**, 417 (1990).
10. W. Fischler, D. Morgan, and J. Polchinski, *Phys. Rev.* **D41**, 2638 (1990).
11. W. Fischler, D. Morgan, and J. Polchinski, *Phys. Rev.* **D42**, 4042 (1990).
12. K. Sato, M. Sasaki, H. Kodama, and A. Maeda, *Prog. Theor. Phys.* **65**, 1443 (1981); H. Kodoma, M. Sasaki, K. Sato, and K. Maeda, *Prog. Theor. Phys.* **66**, 2052 (1981); K. Sato, *Prog. Theor. Phys.* **66**, 2287 (1981); H. Kodama, M. Sasaki, and K. Sato, *Prog. Theor. Phys.* **68**, 1979 (1982); K. Maeda, K. Sato, M. Sasaki, and H. Kodama, *Phys. Lett.* **B108**, 98 (1982); K. Sato, H. Kodama, M. Sasaki, and K. Maeda, *Phys. Lett.* **B108**, 103 (1982).
13. V. A. Berezin, V. A. Kuzmin, and I. I. Tkachev, *Phys. Lett.* **B120**, 91 (1983); *Phys. Rev.* **D36**, 2919 (1987); also in *Proceedings of 3rd Seminar on Quantum Gravity 1984*, eds. M. A. Markov, V. A. Berezin, and V. P. Frolov (World Scientific, Singapore, 1985), p. 605.
14. J. Ipser and P. Sikivie, *Phys. Rev.* **D30**, 712 (1984).
15. A. Aurilia, G. Denardo, F. Legovini, and E. Spallucci, *Phys. Lett.* **B147**, 258 (1984); *Nucl. Phys.* **B252**, 523 (1985).
16. K. Lake, *Phys. Rev.* **D19**, 2847 (1979); K. Lake and R. Wevrick, *Can. J. Phys.* **64**, 165 (1986).
17. S. K. Blau, E. I. Guendelman, and A. H. Guth, *Phys. Rev.* **D35**, 1747 (1987).
18. A. Aurilia, M. Palmer, and E. Spallucci, *Phys. Rev.* **D40**, 2511 (1989).
19. W. Israel, *Nuovo Cimento* **B44**, 1 (1966), and correction in **B48**, 463 (1967).
20. M. D. Kruskal, *Phys. Rev.* **119**, 1743 (1960); G. Szekeres, *Publ. Mat. Debrecen* **7**, 285 (1960).
21. M. Aryal and A. Vilenkin, *Phys. Lett.* **B199**, 351 (1987).
22. A. D. Linde, *Mod. Phys. Lett.* **A1**, 81 (1986); A. S. Goncharov, A. D. Linde, and V. F. Mukhanov, *Int. J. Mod. Phys.* **A2**, 561 (1987).
23. A. A. Starobinsky, *Phys. Lett.* **B117**, 175 (1982); see also A. A. Starobinsky, in *Field Theory, Quantum Gravity and Strings*, eds. M. J. de Vega and N. Sánchez, Lecture Notes in Physics (Springer-Verlag) **246**, 107 (1986).
24. A. H. Guth and S.-Y. Pi, *Phys. Rev.* **D32**, 1899 (1985).
25. E. Farhi and A. H. Guth, *Phys. Lett.* **B183**, 149 (1987).
26. S. Coleman and F. De Luccia, *Phys. Rev.* **D21**, 3305 (1980).
27. A. H. Guth and E. J. Weinberg, *Nucl. Phys.* **B212**, 321 (1983).
28. G. W. Gibbons and S. W. Hawking, *Phys. Rev.* **D15**, 2738 (1977).

THE KLEIN–NISHINA FORMULA

GÖSTA EKSPONG
Department of Physics
Stockholm University
Stockholm, Sweden

During the 1920's the Compton effect was very much at the centre of interest among physicists due to its demonstration of the inadequacy of the classical theory for light when trying to account for the scattering of X-rays. The angular intensity distribution of scattered X-rays had not been successfully described by theory until Oskar Klein and Yoshio Nishina in 1928 solved the problem using the then new Dirac theory for the electron. The present paper gives an account of the collaboration in the summer of 1928 between the two physicists — one a theorist, the other an experimentalist — when both were visitors at the Niels Bohr Institute in Copenhagen.

1. The Klein–Nishina Collaboration

When Dirac paid a short visit to Copenhagen in the spring of 1928, he met Klein and Nishina. The three of them were once conferring in the library of the Niels Bohr Institute. Dirac was a man of few words, so when the remark came from Nishina that he had found an error of sign in the then new Dirac paper on the electron, Dirac drily answered: "But the result is correct." Nishina, in an attempt to be helpful, said: "There must be two mistakes," only to get Dirac's reply that "there must be an even number of mistakes."[1]

The reason behind Nishina's seemingly thorough knowledge of Dirac's paper was that he had been assigned to compute the intensity of the scattered radiation (i.e. the cross section) for Compton scattering with the new electron theory. Walter Gordon, during a short visit to Copenhagen somewhat earlier that spring, had suggested that it would be a suitable problem for the young Japanese physicist. He knew Nishina, who had visited Hamburg to learn theory before returning to Japan. When Gordon made that suggestion, Klein immediately agreed, although he himself had intended to attack the same problem.

Klein had in fact partly treated Compton scattering the preceding year and Gordon had given it a full treatment, both using semi-classical theory.

There was no established method available to solve the problem, which made it difficult. When the time came to leave Copenhagen for the summer vacation, "Nishina had — quite obviously — not been able to do more than a general preparation of the problem", as Klein wrote[1] almost 50 years later. Klein suggested a collaboration during the summer of 1928. Nishina settled in a mansion in Lundeborg near the place on the Danish island Fyn where Klein and his family were renting a summer house.

The two of them worked day by day independently during the summer, doing the lengthy algebra while seated in cosy folding chairs.[2] At day's end they compared notes. In early July 1928 Klein wrote to Bohr informing him of the progress,[3] saying that a first draft was being written. Their joint short letter to *Nature*[4] was signed on the 3rd of August 1928. The full paper had by then not been finished, as told by Nishina in a letter to Bohr.[5] The plan was to finish it in September, when both of them would be back in Copenhagen. Their joint paper was received by *Zeitschrift für Physik* on the 30th of October.[6]

A second paper, with formulae for polarization phenomena, was published immediately after the first one and was signed by Yoshio Nishina alone.[7] Its content is alluded to in the first paper. When it was submitted, Nishina had already been in France for several weeks on his way back to Japan. According to letters between Copenhagen and Tokyo, some errors in the first form of some formulae had been found by Christian Møller in Copenhagen and corrections made to the final paper. As seen by the correspondence and also by an erratum to *Nature*,[8] the error was an oversight of the fact that the photon frequency will be changed during the first scattering in a double scattering process.

2. Who Were the Two Physicists?

Oskar Klein started on chemistry at a very young age, in the laboratory of Svante Arrhenius (Nobel Prize winner in Chemistry in 1903). He obtained his Ph.D. in Stockholm, his home town, where his father was the chief rabbi of the Jewish congregation. On his own and while still at the chemistry laboratory he read the papers on quantum theory by Bohr, Sommerfeld, Einstein and Debye. His other field of study was statistical mechanics. He went in 1918 to Bohr in Copenhagen, when he was 24 years old (being born in 1894). Klein transformed there into a clever, original theoretical physicist while discussing with Bohr and Kramers and working on his own. His name is associated with the first five-dimensional theory, today revived again in other contexts under the term Kaluza–Klein theories. George Uhlenbeck has reminisced: "I remember that in the summer of 1926, when Oskar Klein told us of his ideas which would not only unify the Maxwell with the Einstein equations but also bring in quantum theory, I felt a kind of ecstasy! Now one understands the world!"[9]

Oskar Klein stayed with some interruptions in Copenhagen until 1930, when he became a professor at Stockholm University. Since his life history has been told in this series of publications,[10,11] the reader is referred to those accounts as well as to some new material in this volume.[12]

Yoshio Nishina was trained as an electrical engineer at the Tokyo Imperial University, where he graduated in 1918. He became a research fellow at RIKEN (Research Institute for Physics and Chemistry) and majored in physics at the Graduate School of the Tokyo Imperial University. In 1921, at the age of 31 years (he was born in 1890), he left Japan for Europe. He went to Copenhagen in 1923 to stay at Bohr's institute. His research was devoted to X-ray experiments until he went in November 1927 to Hamburg to learn quantum mechanics under the guidance of Wolfgang Pauli. On his return to Copenhagen in March 1928, he began his work on the theoretical calculation of Compton scattering, together with Oskar Klein.

Nishina left Copenhagen in October 1928 and returned in December of that year to Japan, where he became a member of Nagaoka's laboratory at RIKEN. Over the years Nishina became a leading scientist in Japan, building a new laboratory in 1931, starting nuclear physics research and also studies of cosmic rays in 1935, setting up a small cyclotron in 1937 and a big cyclotron over the years 1938–1944. Leading European physicists came to visit Japan, partly because they knew Nishina from his eight years in Europe.

In August 1945 Nishina went to investigate the sites of the nuclear bombing in Hiroshima and Nagasaki. The occupation army soon destroyed the two cyclotrons. The larger one was dumped at sea and has not been found; the smaller one can be seen in Tokyo outside the building where Nishina had his office and where the Nishina archives are located.

RIKEN was reorganized after the war with Nishina as president of the RIKEN company. He died in January 1951, due to liver cancer, at the age of 61 years.

3. Compton Scattering

In his Nobel lecture in 1927, Arthur Compton emphasized that X-rays share all the properties of ordinary light (i.e. reflection, refraction, diffuse scattering, polarization, diffraction, emission and absorption spectra) and therefore are a wave phenomenon. The particle nature is revealed in the photoelectric effect and in the scattering of X-rays in light elements.

That some of the scattered X-rays are shifted towards longer wavelengths is impossible to understand in classical theory. The precise experiments by Compton[13] in 1922 focused the interest on this new phenomenon. "He found an experimental method that gave results which were as exact as they were astonishing" was said at the Nobel ceremony in 1927 by Professor Manne Siegbahn, himself a Nobel laureate and a member of the Nobel Committee for Physics.[14] The explanation by Compton himself and also by Peter Debye[15] was based on the idea that a light quantum (a photon), carrying momentum as well as energy, collides with a free electron in the scattering material, and that energy and momentum are each conserved in the individual act of collision. Relativistic kinematics predicts that the wavelength shift is independent of the energy of the primary radiation and of the scattering material. It depends on the scattering angle, ϕ, in a simple manner:

$$\Delta\lambda = \frac{h}{mc}(1 - \cos\phi)\,. \tag{1}$$

This formula agreed very well with accurate measurements. This came as a striking confirmation of Einstein's concept of light quanta from 1905 (the word "photon" was coined in 1926). Early in 1923 Sommerfeld wrote to Bohr: "The most interesting thing... is the work of Compton in St. Louis.... After it the wave theory of X-rays will become invalid."[16]

It seems that the 26-year-old H. A. Kramers had already set up the kinematical equations of a photon–electron collision in 1921, before the discovery of the phenomenon. As told by Max Dresden[17] Kramers was thoroughly converted by Bohr to believing that photons do not exist and that he should not publish this. Kramers became one of strongest opponents of the photon concept, even long after the Compton effect had been fully explored experimentally.

Several leading physicists, including Max Planck and Niels Bohr, resisted Einstein's concept of light quanta for a long time. There are quotes from Bohr *before* the discovery of the Compton effect and *after* it, which express his resistance. In 1922 Bohr rejected the usefulness of the concept in his Nobel lecture, saying: "In spite of its heuristic value [Einstein's] hypothesis of light quanta, which is quite irreconcilable with the so-called interference phenomena, is not able to throw light on the nature of radiation."[18] Einstein himself was invited to the same Nobel event, and should certainly have been present were it not for his journey to Japan. He was awarded the 1921 Nobel

Prize (reserved in 1921 and awarded in 1922) on the same occasion as Bohr "...for his discovery of the law of the photoelectric effect." It is a pity that an encounter in December 1922 in Stockholm between the two giants in physics never took place. After the Compton effect had been discovered Bohr was troubled, as is clear from some words in a letter[19] written in January 1924 to Ernest Rutherford: "You can understand my concern — for a person for whom the wave theory is a creed."

To people like Niels Bohr and Max Planck, light was demonstrably a wave motion and Maxwell's theory accounted for it in a convincing way. There existed no technique to describe light as waves and particles at the same time. The quantum phenomena were viewed by them as having nothing to do with light motion in vacuum, only with light interacting with matter — a subject not fully understood anyway. After a last effort to avoid the necessity of the photon concept in a 1924 paper by Bohr, Kramers and Slater, new experiments in 1925 on time and space coincidences between scattered quanta and recoil electrons in the Compton effect convinced Bohr, that the light quantum, i.e. the photon, was an indispensable concept. Kramers, however, was not convinced. Wolfgang Pauli, in a 1925 letter to Kramers,[20] expressed his opinion in the following words: "It can now be considered as proven — to every objective physicist — that photons are as physically real (or as unreal!) as electrons."

However, in 1927 Prof. Manne Siegbahn claimed in his Nobel Prize presentation of Arthur Holly Compton:[14] "The Compton effect has, through the latest evolutions of the atomic theory, got rid of the original explanation based upon a corpuscular theory. The new wave mechanics, in fact, leads as a logical consequence to the mathematical basis of Compton's theory. Thus the effect has gained acceptable connection with other observations in the sphere of radiation." The modern reader is reminded of the fact that in 1927, when the prize award was decided, neither the Dirac field quantization nor the Bohr complementarily ideas were known.

From the contents of the Nobel Archives[21] it is quite clear what the speaker had in mind when claiming that the particle aspect of X-rays would no longer be needed. The Nobel Committee had already assessed the Compton effect in 1925 and 1926 and found its theory very unsatisfactory. However, in 1927 that had changed. The new evaluation was carried out by Carl Wilhelm Oseen, professor in mechanics and mathematical physics at Uppsala. He did a thorough study for the Committee. He begins by recalling the great interest by which Compton's discovery in 1922 had been met, much due to the theory offered by Compton. He writes: "It is not surprising that the agreement of this theory with observations inspired with less critical representatives for theoretical physics the thought that the lengthy fight between the wave theory and the corpuscular theory would be nearing its end. Compton's discovery was by these scientists taken as the decisive proof for the truth of the corpuscular theory. If these expectations had been fulfilled, the discovery by Compton would undoubtedly have marked a decisive turning point in the development of the whole of the radiation theory." Oseen set out to show that this is not so. His view was that the effect is nevertheless very important.

Oseen tells how the Bohr theory had fallen by 1925 and says that the Compton effect had nothing to do with that. He describes how matrix mechanics and wave mechanics had entered the stage without inspiration from the Compton effect. In a short interlude he mentions Compton's suggestion in 1920 that the electron is spinning around its axis with a quantized angular momentum and constitutes the elementary magnet in matter. After mentioning that Uhlenbeck and Goudsmit in 1925 published their similar idea, Oseen points out that the basis is weak, since it presupposes a rotating rigid body, which is relativistically not acceptable.

The oldest theories for the Compton effect referred to by Oseen were given by Compton, "Debijes" (Debye) and Woos. Being based on the theory of light quanta, "they have been of value to experimental research, but must now be considered obsolete in view of the latest theories." Oseen mentions several such newer works, especially that by Gordon and a recent one by O. Klein, based on the wave theory with no mention of corpuscles (photons). They all arrive at the same equations for the conservation of energy and momentum between the scattered wave and the recoiling electron as originally derived by Compton assuming a two-particle collision. "The basis for the Compton–Debijes theory is thus found, this time not as a hypothesis but as a consequence of the atomic theory," is a conclusion by Oseen, which justifies his judgement of the former being obsolete. These wave-mechanical treatments, furthermore, also gave formulae for the intensity (i.e. cross section) in agreement with measurements. Oseen summarizes by saying that the revolution during the last 18 months has been independent of Compton's discovery and that the new direction for the revolution has moved opposite to the one expected after Compton's discovery. The new theory is a wave theory in a higher degree than any previous theory. By the use of the new theory it has been possible to give a qualitatively and quantitatively correct account of the Compton effect.

The Committee follows Oseen in emphasizing that the Compton effect is important, since it once more and very clearly and convincingly demonstrates that the classical theories are not applicable in the realm of atomic physics and that it offers a valuable and welcome possibility of testing the new ideas.

There were thus at that time some recent works by Schrödinger,[22] by Gordon[23] and by Klein,[24] which induced the statement made at the Nobel Ceremony in 1927. In Schrödinger's view a plane light wave would interact with a moving electron, represented by a plane wave serving as a diffraction "grating." The diffracted plane wave X-ray would be emitted in a direction determined by interference. The wave length shift was viewed as a Doppler effect, since the grating was a moving one. The same kinematical formulae were obtained as on the assumption of a particle collision.

During a short stay at Leiden in 1926 Klein and Uhlenbeck began — but never finished — a calculation of the Compton effect based on Schrödinger's unrelativistic wave mechanics. In a paper[24] published in 1927, the same year as when Compton was awarded the Nobel Prize, Klein showed that the interaction between an electromagnetic wave and an electron, described by his own relativistic wave function (often referred to as the Klein–Gordon equation), leads to the same kinematical relations as those derived on the corpuscular theory.

The observed angular distribution of scattered radiation presented a second and more difficult problem. Classical theory predicts a forward–backward symmetric angular distribution. The observations, however, showed a strongly reduced backward scattering intensity and furthermore an integrated total cross section, decreasing with energy and less than the classical "Thomson" cross section. There were early futile attempts to explain the large deviations from what classical theory predicts. Some of these were almost desperate, as can be seen by the collection of published papers, which were reviewed by Compton in his monograph[25] from 1926. Some formulae were judged by Compton to be wrong, since they disagreed either with experiments or with a general limit obtained by applying Bohr's correspondence principle. "An interesting novel solution of the problem has been presented by Jauncey" are the words of Compton, by which he introduced an idea that photons bounce off the surface of a suitably shaped electron. To make the mark the shape would have to be special and to change with the velocity of the electron. It is perhaps not surprising that Compton considered Jauncey's approach interesting, since it was the only one applying the corpuscular theory to the problem.

In the end two formulae were favoured, one given by Compton himself and the other by G. Breit, with some preference for the latter. The one by Breit, partly based on empirical grounds, is

$$\frac{d\sigma}{d\Omega} = \frac{\sigma_0}{2} \frac{1 + \cos^2\phi}{[1 + \alpha(1 - \cos\phi)]^3}, \tag{2}$$

where α is the photon energy in units of the electron rest mass energy, and thus $\alpha = \frac{h\nu}{mc^2}$.

In the forward direction ($\phi = 0$) one gets σ_0, which is the classical Thomson cross section. In the backward direction the classical scattering formula gives the same value, whereas the Breit formula predicts a smaller and energy-dependent cross section, $\frac{d\sigma}{d\Omega}(\phi = \pi) = \frac{\sigma_0}{1+2\alpha}$. The integrated total scattering cross sections are in the two cases hardly distinguishable from each other in the energy range, where most measurements existed in 1926,

$$\text{Compton:} \qquad \sigma = \frac{\sigma_0}{1+2\alpha}, \tag{3}$$

$$\text{Breit:} \qquad \sigma = \sigma_0 \frac{3}{4} \frac{1+\alpha}{\alpha^3} \left[\frac{2\alpha(1+\alpha)}{1+2\alpha} - \ln(1+2\alpha) \right]. \tag{4}$$

However, the monograph was written before any quantum-mechanical calculations had been published. There were two such before the one by Klein and Nishina. Based on the Klein–Gordon relativistic formula for the electron, i.e. an electron without spin, Dirac published a solution[26] in 1926, and Gordon[23] one in 1927. There was also the solution given by Schrödinger,[22] inspired by the Gordon paper. These gave the same result as the one by Breit, Eq. (2) above, for the differential cross section. That was still not the final, correct solution to the problem, since when the electron spin is included, a more complicated formula results, as first shown by Klein and Nishina. The magnetic moment associated with the spin of the electron causes additional scattering, which varies in magnitude with the angle of scattering and also depends on the energy of the photon.

4. On the Derivation of the Klein–Nishina Formula

The full paper[27] by Klein and Nishina, which was published in German, appears translated into English for the first time in this volume. A short note[28] in English was published in 1928. A modern reader will be surprised by the methods used by Klein and Nishina in their derivation of the famous Klein–Nishina formula. One must not forget, however, that no established methods for such problems existed at that time, long before Feynman graphs were invented.

Klein and Nishina took as starting point the then new relativistic theory for the electron by Dirac. They considered the recently published treatments of the problem Dirac and by Gordon as being based on the "older" forms of relativistic quantum mechanics. When it came to handling the Compton scattering problem, Klein used Dirac's relativistic theory, whereas Dirac had used Klein's. But that was before Dirac had found his relativistic theory for the electron.

A modern reader of the Klein–Nishina paper is struck by the fact that Dirac's field quantized theory is not applied, although it was already available. There are thus in their paper no processes of photon absorption by the electron, no intermediate states, no re-emissions of a photon. They did not apply a quantized electromagnetic field, but rather a classical field. Klein was a pioneer in semi-classical theory. The negative energy states as solutions to the Dirac equations were considered physically without meaning. One sentence in the paper reads: "We will of course limit ourselves to positive values." There was also the problem of boundary conditions. They claimed that a treatment according to Dirac's radiation theory would yield the same result as theirs, when one limits oneself to the first approximation. This was later — in 1930 — shown to be true by the detailed calculations of Ivar Waller[29] and Igor Tamm.[30]

The field of the incoming electromagnetic wave was considered to perturb the electron; "we want to regard a free electron illuminated by a continuous monochromatic wave train." The charged current density was set up and treated so that the radiation field at a distant point could be calculated. They considered the electron to make a transition from a state with momentum p to a state with momentum p' and found that the same relation holds for the frequency shift of the scattered radiation and the scattering angle as in the case of particle collisions.

For the second problem — the angular distribution — they calculated the magnetic field at some distant point in a chosen direction. In their wave-mechanical language the angle of scattering is the angle between the direction of observation and the normal to the wave front of the incoming plane wave. "After some lengthy calculations" they found an expression for the squared field strength in dependence of the scattering angle, the polarization vector and natural constants (e, m, c). Photons were introduced towards the end. Dividing the scattered energy by the photon energy, they obtained the number of scattered photons. The scattering cross section was found to contain one factor in addition to those already given by the Gordon and Dirac formula[23,26] which was reproduced in Eq. (2) above. This new factor,

$$1 + \frac{\alpha^2(1-\cos\phi)^2}{(1+\cos^2\phi)[1+\alpha(1-\cos\phi)]}\,, \tag{5}$$

where $\alpha = \frac{h\nu}{mc^2}$, becomes approximately unity if the radiation frequency, i.e. the photon energy is not too large. Thus the Klein–Nishina formula reduces to the

Dirac–Gordon formula at moderate photon energies and all of them have the energy-independent classical Thomson formula,

$$\frac{d\sigma}{d\Omega} = \frac{e^4}{2m^2c^4}(1+\cos^2\phi)\,, \tag{6}$$

as the low energy limit.

The total cross section, obtained by integration over angles, is

$$\sigma = \sigma_0 \frac{3}{4}\left\{\frac{1+\alpha}{\alpha^2}\left[\frac{2\alpha(1+\alpha)}{1+2\alpha} - \ln(1+2\alpha)\right] + \frac{1}{2\alpha}\ln(1+2\alpha) - \frac{1+3\alpha}{(1+2\alpha)^2}\right\}, \tag{7}$$

where $\alpha = \frac{h\nu}{mc^2}$.

This formula was found to be in very good agreement with Chao's data using γ-rays and with Read and Lauritsen's on X-ray measurements, as stated in 1934 by Compton and Allison in their well-known monograph.[31] It was clearly superior to the formula by Breit, Dirac and Gordon, as displayed in Eq. (4).

There exist manuscripts with unpublished details of the lengthy calculations by Klein and Nishina in the Nishina archives in Tokyo. These have recently been investigated by Yuji Yazaki, who published his findings in the *Journal of the History of Science, Japan* (in Japanese, with a summary in English).[32]

In a subsequent paper, printed immediately after the first one, Nishina derived formulae for polarization phenomena.[33] He pointed out that the results clearly differ from any earlier ones. The scattering of linearly polarized light on polarized electrons was shown to result in two incoherent, elliptically polarized scattered rays. This has to do with no spin-flip and spin-flip transitions, respectively. The initial spin state of the electron was found to be irrelevant, as mentioned already in the joint paper. Since the usual geometry in experiments on polarization was scattering twice at right angles, Nishina treated this case theoretically. For this he handled the case of an incoming, elliptically polarized wave. In that case the result was clearly dependent on the initial spin state of the electron. Although Nishina averaged over the initial spin directions, one can see here, for the first time, that Compton scattering against polarized electrons is sensitive to the circular polarization of γ-rays. In practical applications this has turned out to be of great value. A complete theoretical treatment of the general case was given in 1938 by W. Franz,[34] who derived expressions for the scattering of elliptically polarized light against electrons with polarization in an arbitrary direction. Following Gunst and Page[35] one often writes the differential cross section for Compton scattering on polarized electrons as a sum of two terms,

$$\frac{d\sigma}{d\Omega} = \frac{d\sigma_{\mathrm{KN}}}{d\Omega} + P\,\frac{d\sigma_1}{d\Omega}\,,$$

where the first term is the usual Klein–Nishina expression for unpolarized or linearly polarized light. The second term, which may add or substract, depends on the circular polarization P of the photons.

5. The Negative Energy States in 1928–30

According to a recent study of available archival material[32] in Tokyo, the greatest problem for Klein and Nishina was the treatment of Dirac spin states and how to interpret them in a physically meaningful way. There were no established methods. Klein devised a method to derive the states from solutions of electrons at rest by a Lorentz transformation.

In their paper Klein and Nishina avoided explicit reference to electron states with negative energy. However, these did appear somewhat later when the Compton scattering problem was treated with quantized fields according to Dirac's radiation theory. The first to do so was Ivar Waller in Uppsala, who published[29] that in 1930. A similar treatment[30] came at about the same time from Igor Tamm in Moscow. Both of them used the original Dirac theory for the electron with empty negative energy states. Both found that only if these states were not ignored did one arrive at the Klein–Nishina formula. Waller calculated sums over positive energy states and negative energy states separately and claimed that both were needed. Tamm was even more expressive, in saying that the result would be grossly wrong if the negative energy states were left out. Both of these authors found furthermore the surprising result that in the classical limit of low photon energies, scattering via intermediate states with positive energies gives zero contribution, and that the Thomson cross section derives from scattering via negative energy states alone. Thus the Klein–Nishina formula came to serve as a test when one was clarifying the importance of the ill-understood negative energy states.

Dirac heard of Waller's result before publication and immediately created the idea of filling the negative energy sea with unobservable electrons. In this new form of his theory there would be a replacement of transitions to now-no-longer available negative energy states by transitions of a new type, in which a hole state appears (a virtual positron in modern terminology). When Dirac described the consequences of his hole theory in a letter[36,37] to Bohr, his words were almost as clear as a Feynman diagram for the Compton scattering: "On my new theory ... there is ... a new kind of double transition taking place in which first one of the negative-energy electrons jumps to the proper final state with emission (absorption) of a photon, and secondly the original positive-energy electron jumps down and fills up the hole with absorption (emission) of a photon. This new kind of process just makes up for those excluded and restores the validity of the scattering formulas, derived on the assumption of the possibility of intermediate states of negative energy." Waller, who knew about Dirac's new hole theory before that was published, claimed in his paper that the result would be the same, since only formal changes would be needed. Shortly afterwards, Dirac proved this to be true.[38]

In view of the modern interpretation of negative energy states as virtual positrons and with regard to the results by Waller and Tamm, one reaches the not-so-obvious conclusion that scattering of light in the classical limit proceeds through intermediates states with three charged particles present, one of which is a virtual positron, and the other two the incoming and the outgoing electron.[39]

6. A New Tool for Research

6.1. *Cosmic rays*

The nature and energy of cosmic rays was under debate and investigation in the late 1920's. Some believed they are hard γ-rays. Klein and Nishina suggested that their formula could serve as a tool for determining the energies of cosmic rays, since the cross section is a steeply decreasing function of increasing γ-ray energy, E_γ,

$$\sigma \approx \frac{\pi\alpha^2}{mE_\gamma} \ln \frac{2E_\gamma}{m} ,$$

where α now is the fine structure constant.

It follows that if there were no other processes entering at high energy, cosmic γ-rays would be very penetrating and more so the higher the energy. It will be recalled that Nishina, when back in Tokyo, started a programme of research in cosmic rays.

However, the scheme never worked as envisaged, since another new process enters at energies larger than about 1 MeV. This process is, of course, pair creation, which reveals the real meaning of the mysterious, negative energy states.

6.2. *Positrons*

When experiments were made to test the correctness of the Klein–Nishina formula, it was found that filters of high-Z materials attenuated high energy

γ-rays much more than predicted, but that the attenuation in low-Z materials was as expected.[40–42] Disagreements with the formula were taken as hints that some new process exists, the nature of which was missed, however, by the early researchers and by others as well.[43] The new process was identified with the creation of electron–positron pairs (which occurs much more readily in high-Z materials) only much later, after the discovery of the free positron by Carl David Anderson in 1932 and further studies by Blackett and Occhialini.

Still another new phenomenon was discovered in 1930 by Chung-Yao Chao, a Chinese visitor to California Institute of Technology, who also had been one of those observing the discrepancy referred to above.[42] He reported the existence of a new kind of secondary radiation coming from a heavy target irradiated by high energy γ-rays.[44] Chao described the situation in the following way: "I first found that the angular distribution of the secondary radiation produced by an Al-scatterer agrees well with the Klein–Nishina formula. But in the case of Pb, in addition to the Compton scattering, I found a special radiation which is very prominent in the backward direction where the intensity of Compton scattering is very weak. I estimated the photon energy of this special radiation by using the absorption method with Pb plates and found it to be about 550 keV." The radiation was emitted roughly isotropically and was of such an energy that with hindsight it was identified with e^+e^- annihilation radiation. Thus the Klein–Nishina formula served as a guide to making new discoveries — close to a discovery of positrons — but not quite making the mark.

6.3. *Polarization of annihilation radiation*

In two-photon annihilations of electron–positron pairs, which is related to Compton scattering by crossing, theory[45] predicts polarization correlations, which can be described either as perpendicular correlations of their linear polarizations or as both photons having the same helicity, when their circular polarizations are analyzed. The first type of correlation has been observed in the azimuthal variation of the coincidence rate of Compton-scattered annihilation quanta.[46] Evidence for the second type was claimed in another experiment.[47]

6.4. *The helicity of the neutrino*

In a very beautiful experiment on the helicity of the electron neutrino,[48] M. Goldhaber, L. Grodzins and A. W. Sunyar arranged so that the neutrino helicity in an electron capture process was transferred to the circular polarization of the subsequently emitted γ-ray. They used the sensitivity of Compton scattering to the state of circular polarization of the photons when scattered against polarized electrons. The transmission of these γ-rays through a magnetized iron filter therefore depends on the direction of the magnetized field relative to the line of flight of the photons. The main ingredients in the experiment were given a concise summary by the authors: "A combined analysis of circular polarization and resonant scattering of γ-rays following orbital electron capture measures the helicity of the neutrino." The result was that the neutrino emitted in beta decay has negative helicity, which at that time added strength to the hypothesis of the V–A structure of weak interactions.

The results of the collaboration between Klein and Nishina in 1928 thus came to play an important role in the study of parity violations and in determinating the nature of the weak interactions 30 years later.

6.5. *Applications in other research*

There are numerous other applications where the results of the collaboration between Klein and Nishina are used. In nuclear and particle physics there are many applications, in particular to determine the polarization state of γ-rays. In astrophysics it has played a role somewhat different from the one envisaged by the authors in 1928, namely one considers inverse Compton scattering, a process in which energy is transferred from a very fast electron to a low energy photon, making it a photon of very high energy. Recently discovered astrophysical point sources emitting photons in the TeV energy range may depend on such processes. The cross section for such an energy transfer is basically obtained from the Klein–Nishina formula with suitable Lorentz transformations, since the formula holds for an electron initially at rest. High energy photon beams have been created in the laboratory by backward scattering of laser light from a beam of high energy electrons. Compton scattering has also been applied to problems in solid state physics, for which very high energy (10 GeV) electron storage rings, used as synchroton facilities, are expected to become useful.

Acknowledgements

Informative and penetrating discussions with Dr. Lars Bergström (Stockholm) have been very valuable and are acknowledged. I would like to express my thanks for useful and informative discussions with Prof. Chen Ning Yang (Stony Brook, USA). Permission was granted to use the Nobel Archives by the Royal Swedish Academy of Sciences. I am indebted to Dr. Finnn Aaserud (Copenhagen) for his efforts and kindness in searching the Niels Bohr archives for pertinent material; and to Drs. Tamaki, Takeuchi and Yazaki (Tokyo) for showing me relevant material in the archives of the Nishina Memorial Foundation in Tokyo, Dr. Yazaki has sent me his recent study of the Klein–Nishina collaboration with comments in English, for which I am deeply grateful.

References

1. "Oskar Klein: To the Memory of Yoshio Nishina", manuscript (1975), the archives of the Nishina Memorial Foundation, Tokyo.
2. B. Laurent, private communication.
3. O. Klein, letter to N. Bohr, July 6, 1928, The Niels Bohr Archives BSC (13,1).
4. O. Klein and Y. Nishina, *Nature* **122**, 398 (1928).
5. Y. Nishina, letter to N. Bohr, August 8, 1928, The Niels Bohr Archives BSC (14,2).
6. O. Klein and Y. Nishina, *Z. Phys.* **52**, 853 (1929).
7. Y. Nishina, *Z. Phys.* **52**, (1930) 869.
8. Y. Nishina, *Nature* **123**, 349 (1929).
9. Abraham Pais, *Subtle Is the Lord* (Oxford University Press, 1982), p. 332.
10. Inga Fisher-Hjalmars and Bertel Laurent, "Oskar Klein", in Vol. 1 of *The Oskar Klein Memorial Lectures* (World Scientific, Singapore, 1991).
11. Oskar Klein, "From My Life of Physics", *ibid.*, p. 103.
12. O. Klein, "Excerpts from Some Autobiographical Notes" (transl. from Swedish), this volume.
13. A. H. Compton, *Phys. Rev.* **21**, 483 (1923).
14. *Les Prix Nobel en 1927* (P. A. Norstedt & Söner, Stockholm, 1928), p. 19.
15. P. Debye, *Phys. Z.* **24**, 161 (1923).
16. Letter from Sommerfeld to Bohr, January 21, 1923 (*Archives for the History of Quantum Physics*).
17. Max Dresden, *H. A. Kramers: Between Tradition and Revolution* (Springer-Verlag, 1987).
18. *Les Prix Nobel en 1921–1922* (P. A. Norstedt & Söner, Stockholm, 1923).
19. N. Bohr, Letter to Ernest Rutherford, January 9, 1924 (*Archives for the History of Quantum Physics*).
20. W. Pauli, Letter to H. A. Kramers, July 27, 1925 (*N. Bohr Collected Works*, Vol. 5, p. 442).
21. The Nobel Archives, the Royal Swedish Academy of Sciences, Stockholm: Report 1927 by the Nobel Committee for Physics (excerpts transl. from the Swedish by the author).
22. E. Schrödinger, *Ann. Phys.* **82**, 257 (1927).
23. W. Gordon, *Z. Phys.* **40**, 117 (1927).
24. O. Klein, *Z. Phys.* **41**, 407 (1927).
25. A. H. Compton, *X-Rays and Electrons.*, 1st ed., pp. 296–314 (Van Nostrand, 1926).
26. P. A. M. Dirac, *Proc. R. Soc.* **A111**, 405 (1926).
27. O. Klein and Y. Nishina, *Z. Phys.* **52**, 853 (1929).

28. O. Klein and Y. Nishina, *Nature* **122**, 398 (1928).
29. I. Waller, *Z. Phys.* **61**, 837 (1930).
30. I. Tamm, *Z. Phys.* **62**, 545 (1930).
31. A. H. Compton and S. K. Allison, *X-Rays in Theory and Experiment* (Van Nostrand, New York, 1935).
32. Y. Yazaki (*Kagakusi Kenkyu, J. Hist. Sci. Japan*), II, 31, part I, p. 81; part II, p. 129 (1992).
33. Y. Nishina, *Z. Phys.* **52**, 869 (1930).
34. W. Franz, *Ann. Phys.* **33**, 47 (1938).
35. S. B. Gunst and L. A. Page, *Phys. Rev.* **92**, 970 (1953).
36. P. A. M. Dirac, Letter to N. Bohr, November 1929, quoted in Ref. 37, p. 350.
37. A. Pais, *Inward Bound*, 1st ed. (Oxford University Press).
38. P. A. M. Dirac, *Proc. Cambridge Philos. Soc.* **26**, 361 (1930).
39. Cf. e.g.: J. D. Bjorken and S. D. Drell, *Relativistic Quantum Mechanics* (McGraw-Hill, 1964), pp. 10–12.
40. G. T. P. Tarrant, *Proc. R. Soc.* **A139**, 345 (1930).
41. L. Meitner and H. H. Hupfeld, *Naturwissenschaften* **18**, 534 (1930) and *Z. Phys.* **67**, 147 (1930).
42. C. Y. Chao, *Proc. Natl. Acad. Sci.* **16**, 431 (1930).
43. Cf. Bing An Li and C. N. Yang, *Int. J. Mod. Phys.* **A4**, 4325 (1989).
44. C. Y. Chao, *Phys. Rev.* **36**, 1519 (1930).
45. J. A. Wheeler, *Ann. N. Y. Acad. Sci.* **48**, 219 (1946); C. N. Yang, *Phys. Rev.* **77**, 136 (1950).
46. R. C. Hanna, *Nature* **132**, 332 (1948); E. Bleuler and H. L. Bradt, *Phys. Rev.* **73**, 1398 (1948); C. S. Wu and I. Shaknov, *Phys. Rev.* **77**, 136 (1950).
47. F. P. Clay and F. L. Hereford, *Phys. Rev.* **85**, 675 (1952).
48. M. Goldhaber, L. Grodzins and A. W. Sunyar, *Phys. Rev.* **109**, 1015 (1958).

ON THE SCATTERING OF RADIATION BY FREE ELECTRONS ACCORDING TO DIRAC'S NEW RELATIVISTIC QUANTUM DYNAMICS*

O. KLEIN and Y. NISHINA

in Copenhagen

Received October 30, 1928

The intensity of Compton scattering radiation is calculated on the basis of the new relativistic quantum dynamics developed by Dirac. The result shows deviations from the corresponding Dirac–Gordon formulae of second order with respect to the ratio of the energy of the primary light quantum to the rest energy of the electron.

1. Introduction

On the foundation of the older form of relativistic quantum mechanics, Dirac[1] and Gordon[2] have developed a theory for the intensity and polarization of Compton scattering radiation, which seems to be in good agreement with experience for not-too-short wave radiation. According to the new relativistic quantum mechanics recently developed by Dirac,[3] in which the phenomena connected with the self-rotation of the electron are automatically taken into account, the foundation of a theory of the scattering of light on free electrons has changed and one can expect that also the final results of the Dirac–Gordon theory for the Compton effect will be influenced by this. In the present work, we have tried to attack the problem of scattering radiation on free electrons on the basis of Dirac's new dynamics of the electron. In this we have followed the treatment given by Gordon, which rests on a correspondence-wise utilization of wave mechanics. One would expect that the radiation theory given by Dirac, which allows one to take radiation damping into account, gives

*Original in *Z. Phys.* **52**, 853 (1929); reproduced here with permission from Springer-Verlag. Translated from the German by Dr Lars Bergström.

a coinciding result in this case when dealing with the first approximation with respect to the intensity of the primary radiation.

In the present work we have restricted ourselves to the calculation of the intensity of the scattered radiation and its dependence on direction and wave length. The question of the polarization of the scattered radiation will be treated by one of us in an accompanying work.[4] It has appeared that in the region of hard γ-rays the deviations of our results from the Dirac–Gordon formulae are substantial so that, for example, the determination of the wave-length of the cosmic penetrating radiation according to the present theory would give considerably shorter wavelengths than in the old theory. But just in this region the experimental results on the Compton effect seem to be too uncertain to give a decision for or against the theory at this time. Here an accurate experimental test would, however, be very desirable, not least in view of the difficulty of his new theory emphasized by Dirac, connected with the possibility of negative energies.

2. Introductory Remarks on the Wave Equation of Dirac

According to Dirac[5] the quantum-mechanical problem of an electron with charge $-e$ and rest mass m, moving in a field of force where the electrostatic potential is V and the vector potential is $\mathcal{U}$ is determined by the following Hamiltonian form F:

$$F = \frac{E + eV}{c} + \rho_1 \left(\sigma, p + \frac{e}{c}\mathcal{U} \right) + \rho_3 mc \,, \tag{1}$$

where E means the energy of the electron and p its momentum vector with the components p_1, p_2, p_3 with respect to the axes x_1, x_2, x_3 of an orthogonal coordinate system, while c denotes the vacuum velocity of light. Further, σ is a matrix vector with components $\sigma_1, \sigma_2, \sigma_3$, fulfilling the relations

$$\left.\begin{aligned} \sigma_1\sigma_2 = i\sigma_3 = -\sigma_2\sigma_1, \quad \sigma_1^2 = 1 \,, \\ \sigma_2\sigma_3 = i\sigma_1 = -\sigma_3\sigma_2, \quad \sigma_2^2 = 1 \,, \\ \sigma_3\sigma_1 = i\sigma_2 = -\sigma_1\sigma_3, \quad \sigma_3^2 = 1 \,, \end{aligned}\right\} \tag{2}$$

and ρ_1 and ρ_3 are two out of three matrices that fulfil relations identical to (2) and moreover commute with σ. As Dirac has shown, all these quantities can be represented by matrices with four rows and columns. Correspondingly, the mutually adjoint eigenfunctions φ and ψ belonging to the Hamiltonian form F, which we will regard as functions of the coordinates x_1, x_2, x_3 and the time t, consist each of four components, $\varphi_1, \varphi_2, \varphi_3, \varphi_4$ and $\psi_1, \psi_2, \psi_3, \psi_4$, respectively, so that an expression like $\mu\psi$, where μ means a four-row matrix, is to be understood as an abbreviation for the four quantities $\sum_{k=1}^{4} \mu_{ik}\psi_k$ ($i = 1, 2, 3, 4$), while $\varphi\mu$ denotes the quantities $\sum_{k=1}^{4} \varphi_k \mu_{ki}$ ($i = 1, 2, 3, 4$), where μ_{ik} are the matrix elements of μ.

The physical utilization of Eq. (1) depends on the assumption that if φ and ψ are associated wave functions of the Hamiltonian function (1), namely that φ and ψ are the complex conjugates of each other for Hermitian matrices ρ_1, ρ_3 and σ, then $\varphi\psi d\tau$ gives the probability that the electron is found in the volume element $d\tau$. From this follows, according to Dirac, for the probability that the electron in a time dt crosses a surface element df, $-c\varphi\rho_1\sigma_s\psi df dt$, where σ_s denotes the component of the vector σ in the direction normal to the surface element, and where $\varphi\mu\psi$, if μ again stands for an arbitrary four-row matrix, should mean $\sum_{ik} \varphi_i \mu_{ik} \psi_k$. If we multiply the quantities $\varphi\psi$ and $-c\varphi\rho_1\sigma\psi$ by the charge $-e$ of the electron, we get the so-called wave-mechanical electric density ρ and the electric current density vector $\mathcal{J}$. Thus

$$\rho = -e\varphi\psi, \quad \mathcal{J} = ec\varphi\rho_1\sigma\psi \tag{3}$$

are valid.

3. Eigenfunctions of the Dirac Wave Equation for a Free Electron

First we want to consider a free electron where we may choose as eigenfunctions, like in the Schrödinger theory, the following, to each value of p associated functions φ_0 and ψ_0, namely

$$\varphi_0(p) = u(p)e^{\frac{i}{h}[Et-(pr)]}, \quad \psi_0(p) = v(p)e^{-\frac{i}{h}[Et-(pr)]}, \tag{4}$$

where h denotes the Planckian constant divided by 2π, and r the coordinate vector with components x_1, x_2, x_3, while $u(p)$ and $v(p)$ are quantities independent of time and coordinates, consisting of four components and having to fulfil the following purely algebraic conditions:

$$\left.\begin{aligned} u(p)\{E/c + \rho_1(\sigma p) + \rho_3 mc\} &= 0\,, \\ \{E/c + \rho_1(\sigma p) + \rho_3 mc\}v(p) &= 0\,. \end{aligned}\right\} \tag{5}$$

To a given p there belong, according to these equations, two values of the energy E, one positive and one negative, which both fulfil the relation

$$E^2/c^2 = m^2c^2 + p^2\,, \tag{6}$$

which expresses the relation between energy and momentum in the relativistic mechanics. As Eqs. (5) describe abbreviations for systems of four equations each, one should really expect that to a given value of p there belong four energy values, which just corresponds to the possibility of taking into account, besides the physically non-sensible negative energy values, the doubling of the number of eigenvalues connected with the self-magnetism of the electron. This degeneracy of the free electron has as a consequence the fact that, if we choose a particular eigenvalue—naturally we will then restrict ourselves to positive values—the ratios of the components of $u(p)$ and likewise of $v(p)$ are not yet fixed but rather depend on two quantities to be chosen freely. In the following calculation of the intensity of the scattered radiation, it is necessary to take this degeneracy into account.

We first want to consider the case of an electron at rest. Here $p = 0$, and we put $E = mc^2$. Equations (5) thus simply become

$$u(0)(1+\rho_3) = 0, \quad (1+\rho_3)v(0) = 0\,. \tag{7}$$

We will choose ρ_3 to be a diagonal matrix, as we put[a]

$$\rho_3 = \begin{pmatrix} -1 & 0 & 0 & 0 \\ 0 & -1 & 0 & 0 \\ 0 & 0 & 1 & 0 \\ 0 & 0 & 0 & 1 \end{pmatrix}. \tag{8}$$

It then follows that Eqs. (7) are satisfied if u_3, v_3, u_4, v_4 are equal to zero, while u_1, v_1 and u_2, v_2 can be chosen freely. We can in other words compose the general solution belonging to the given eigenvalue from two independent solutions for which only u_1, v_1 and u_2, v_2 are different from zero, respectively.

We will now further choose the component σ_3 of σ along the x_3 axis to be a diagonal matrix, as we put

$$\sigma_3 = \begin{pmatrix} 1 & 0 & 0 & 0 \\ 0 & -1 & 0 & 0 \\ 0 & 0 & 1 & 0 \\ 0 & 0 & 0 & -1 \end{pmatrix}. \tag{9}$$

According to Dirac $u(p)\sigma v(p)$ is now proportional to the magnetic moment of the electron belonging to the solution $u(p)$, $v(p)$. The two solutions mentioned thus give an equal but oppositely directed magnetic moment of the

[a]We have, for ρ_3, chosen the opposite sign to that of Dirac, so that for positive energy the two first components of u and v rather than the two last components remain finite.

electron along the x_3 axis. We can therefore regard u_1, v_1 and u_2, v_2, respectively, as the eigenfunctions belonging to the two possible orientations of the magnetic moment with respect to the x_3 axis.

We can easily reduce the case of an arbitrary p-value to the case $p = 0$ by carrying out a contact transformation closely related to the Lorentz transformation.[b] Let

$$S = \alpha + i\beta\rho_2(\sigma p), \quad S^{-1} = \alpha - i\beta\rho_2(\sigma p), \quad \alpha^2 + \beta^2 p^2 = 1, \tag{10}$$

where we soon want to determine α and β closer. We put

$$F^* = S(E/c + \rho_1(\sigma p) + \rho_3 mc)S^{-1}. \tag{11}$$

Now one has

$$S\rho_1 S^{-1} = \rho_1(\alpha - i\beta\rho_2(\sigma p))^2 = (\alpha^2 - \beta^2 p^2)\rho_1 + 2\alpha\beta\rho_2(\sigma p)$$

and thus

$$S\rho_1(\sigma p)S^{-1} = (\alpha^2 - \beta^2 p^2)\rho_1(\sigma p) + 2\alpha\beta p^2 \rho_3 .$$

Further

$$S\rho_3 S^{-1} = (\alpha^2 - \beta^2 p^2)\rho_3 - 2\alpha\beta\rho_1(\sigma p) .$$

It thus follows that

$$F^* = E/c + (\alpha^2 - \beta^2 p^2 - 2\alpha\beta mc)\rho_1(\sigma p) + ((\alpha^2 - \beta^2 p^2)mc + 2\alpha\beta p^2)\rho_3 .$$

We now want to choose α and β such that the coefficient of $\rho_1(\sigma p)$ in F^* vanishes, i.e. we demand

$$\alpha^2 - \beta^2 p^2 - 2\alpha\beta mc = 0 . \tag{12}$$

Further, we set

$$(\alpha^2 - \beta^2 p^2)mc + 2\alpha\beta p^2 = m^* c \tag{13}$$

[b]The following consideration is in close relation to Dirac's treatment of the invariance of his equations with respect to Lorentz transformations; *loc. cit.* B, p. 615.

and then have

$$F^* = E/c + \rho_3 m^* c\,. \tag{14}$$

It now follows from a simple calculation that

$$m^{*2}c^2 = m^2c^2 + p^2\,, \tag{15}$$

and if, in accordance with this, we set

$$m^* c = +\sqrt{m^2c^2 + p^2}\,, \tag{16}$$

we easily get

$$\alpha = \sqrt{\frac{m^* + m}{2m^*}}, \quad \beta = \sqrt{\frac{m^* - m}{2m^* p^2}}\,, \tag{17}$$

as a result of which the condition (12) is fulfilled.

If we now set

$$u(p) = u^*(p)S(p), \quad v(p) = S^{-1}(p)v^*(p)\,, \tag{18}$$

where $S(p)$ and $S^{-1}(p)$ mean the quantities (10) for a definite value of p, it follows from (11) that u^* and v^* satisfy the equations

$$u^*(E/c + \rho_3 m^* c) = 0, \quad (E/c + \rho_3 m^* c)v^* = 0$$

or, since according to (16) $E = m^* c^2$,

$$u^*(1 + \rho_3) = 0, \quad (1 + \rho_3)v^* = 0\,,$$

i.e. we obtain Eqs. (7) for the case $p = 0$. In this manner we can describe $u(p)$ and $v(p)$ with the help of the independent quantities u_1^*, v_1^* and u_2^*, v_2^*, which, however, are less easily connected with the magnetization of the electron than for the case $p = 0$.

Let there now for the moment be a solution $u(p)$, $v(p)$ which corresponds to one of the independent eigenfunctions, where thus either only u_1^*, v_1^* or u_2^*, v_2^* are different from zero. We want to normalize this solution such that it corresponds to one electron. If $\varphi_0(p)$ and $\psi_0(p)$ are the accompanying eigenfunctions in coordinate space, this means, as is known for continuous eigenvalues, that

$$\int \varphi_0(p')\psi_0(p'')dr = \delta(p' - p'')\,, \tag{19}$$

where dr denotes the volume element and the integration is to be extended over the entire space in question, while $\delta(p' - p'')$ means the singular function introduced by Dirac, the integral of which, with respect to p' over an arbitrary region in momentum space including the point $p' = p''$, is equal to 1. By insertion of the expressions (4) for φ_0 and ψ_0 in (19) one obtains, by comparison with the Fourier integral theorem,

$$u(p)v(p) = (2\pi h)^{-3} . \tag{20}$$

Finally, we want to describe the complex conjugate quantities u_1^* and v_1^* and u_2^* and v_2^*, respectively, which belong to the two different eigensolutions $u(p)$, $v(p)$ in the following way by real quantities:

$$\left.\begin{aligned} u_1^*(p) &= a_1 e^{i\delta_1(p)} , \quad & u_2^*(p) &= a_2 e^{i\delta_2(p)} , \\ v_1^*(p) &= a_1 e^{-i\delta_1(p)} , \quad & v_2^*(p) &= a_2 e^{-i\delta_2(p)} . \end{aligned}\right\} \tag{21}$$

According to (20) $a_1^2 = a_2^2 = (2\pi h)^{-3}$ is then valid, so that we can choose the amplitudes a_1 and a_2 such that they are independent of the p value. On the other hand, the phases $\delta_1(p)$ and $\delta_2(p)$ can be chosen completely freely for every p value.

4. Solution of the Wave Equation for an Electron in a Continuous Monochromatic Radiation Field

Here we want to regard a free electron illuminated by a continuous monochromatic wave train. This we can describe through a vector potential $\mathcal{U}$ of the form

$$\mathcal{U} = a e^{i\nu\left(t - \frac{(nr)}{c}\right)} + \bar{a} e^{-i\nu\left(t - \frac{(nr)}{c}\right)} , \tag{22}$$

where a and $\bar{a}$ denote two constant-to-each-other complex conjugate vectors perpendicular to the unit vector n, which gives the direction of propagation of the waves. The position where the potential $\mathcal{U}$ is measured at time t is in this connection characterized through the radius vector r drawn from a fixed point to the position in question. Further, ν means the wave number of the radiation multiplied by 2π. While neglecting higher powers than the first in $|\mathcal{U}|$, we want to search for solutions to the wave equation belonging to F of the form

$$\left.\begin{aligned} \varphi(p) &= \varphi_0(p)\left\{1 + f(p) e^{i\nu\left(t - \frac{(nr)}{c}\right)} - \bar{f}(p) e^{-i\nu\left(t - \frac{(nr)}{c}\right)}\right\} , \\ \psi(p) &= \left\{1 + g(p) e^{i\nu\left(t - \frac{(nr)}{c}\right)} + \bar{g}(p) e^{-i\nu\left(t - \frac{(nr)}{c}\right)}\right\}\psi_0(p) , \end{aligned}\right\} \tag{23}$$

where $\varphi_0(p)$, $\bar{\psi}_0(p)$ are the eigenfunctions of free electrons given in (4), while f, $\bar{f}$ and g, $\bar{g}$ are constant four-row matrices. The general solution of the wave equation follows in the approximation considered by superposition of all possible solutions of the form (23). To determine the quantities f, $\bar{f}$, g, $\bar{g}$ it is useful to start from the equations of second order which according to Dirac can be easily derived from the first order equations belonging to F; they read

$$\left.\begin{aligned}\left\{\frac{h^2}{c^2}\frac{\partial^2}{\partial t^2}+\left(-ih\nabla+\frac{e}{c}\mathcal{U}\right)^2+m^2c^2\right\}\psi+\frac{eh}{c}(\sigma\mathcal{H})\psi+\frac{ieh}{c}\rho_1(\sigma\mathcal{E})\psi=0,\\ \left\{\frac{h^2}{c^2}\frac{\partial^2}{\partial t^2}+\left(ih\nabla+\frac{e}{c}\mathcal{U}\right)^2+m^2c^2\right\}\varphi-\frac{eh}{c}\varphi(\sigma\mathcal{H})+\frac{ieh}{c}\varphi\rho_1(\sigma\mathcal{E})=0,\end{aligned}\right\} \tag{24}$$

where ∇ means the vector operator with components $\frac{\partial}{\partial x_1}, \frac{\partial}{\partial x_2}, \frac{\partial}{\partial x_3}$, while $\mathcal{E}$ and $\mathcal{H}$ are, respectively, the electric and magnetic field strengths belonging to $\mathcal{U}$. By insertion of the expression (23) in (24) one finds that

$$\left.\begin{aligned}f(p)&=\frac{e}{2h\nu(E/c-(np))}(2(ap)+h(\sigma\eta)-ih\rho_1(\sigma\varepsilon)\}\,,\\ \bar{f}(p)&=-\frac{e}{2h\nu(E/c-(np))}\{2(\bar{a}p)+h(\sigma\bar{\eta})-ih\rho_1(\sigma\bar{\varepsilon})\}\,,\\ g(p)&=-\frac{e}{2h\nu(E/c-(np))}\{2(ap)+h(\sigma\eta)+ih\rho_1(\sigma\varepsilon)\,,\\ \bar{g}(p)&=\frac{e}{2h\nu(E/c-(np))}\{2(\bar{a}p)+h(\sigma\bar{\eta})+ih(\sigma\bar{\varepsilon})\}\,,\end{aligned}\right\} \tag{25}$$

where ϵ and $\bar{\epsilon}$ and η and $\bar{\eta}$, respectively, are connected in the following way with the electric field strength $\mathcal{E}$ and the magnetic strength $\mathcal{H}$, respectively, of the radiation field:

$$\left.\begin{aligned}\mathcal{E}&=\varepsilon e^{i\nu\left(t-\frac{(nr)}{c}\right)}+\bar{\varepsilon}e^{-i\nu\left(t-\frac{(nr)}{c}\right)}\,,\\ \mathcal{H}&=\eta e^{i\nu\left(t-\frac{(nr)}{c}\right)}+\bar{\eta}e^{-i\nu\left(t-\frac{(nr)}{c}\right)}\,;\end{aligned}\right\} \tag{26}$$

since

$$\mathcal{E}=-\frac{1}{c}\frac{\partial\mathcal{U}}{\partial t}\,,\qquad \mathcal{H}=\operatorname{rot}\mathcal{U}\,,$$

it follows that

$$\left.\begin{aligned}\varepsilon&=-\frac{i\nu}{c}a\,, &\qquad \bar{\varepsilon}&=\frac{i\nu}{c}\bar{a}\,,\\ \eta&=-\frac{i\nu}{c}[na]\,, &\qquad \bar{\eta}&=\frac{i\nu}{c}[n\bar{a}]\,.\end{aligned}\right\} \tag{27}$$

5. Calculation of the Scattered Radiation Field

We can write the general solution of the wave equations in the presence of the incident radiation in the approximation mentioned in the following way:

$$\Phi = \int \varphi(p)dp\,, \quad \Psi = \int \psi(p)dp\,, \tag{28}$$

where $\varphi(p)$ and $\psi(p)$ denote the approximation solutions, given in (23) and (25), to the corresponding equations — where first no normalization is assumed — while the integration $\int dp$ is to be regarded as an abbreviation for the threefold integration $\int dp_1 \int dp_2 \int dp_3$ over all possible values of the components of momentum p_1, p_2, p_3. The electric current density belonging to the general solution is given by

$$\mathcal{J} = ec\Phi\rho_1\sigma\Psi = ec\int\int \varphi(p)\rho_1\sigma\psi(p')dpdp'\,. \tag{29}$$

In this expression we now want to insert the expressions (23) for φ and ψ and order it in a correspondence-wise way according to the different possible radiation processes.[6] By neglecting quantities of the size of order $|a|^2$ we get, through a simple transformation,

$$\begin{aligned}\mathcal{J} = \mathcal{J}_0 + ce\int\int dpdp'\Big\{&u(p)[\rho_1\sigma g(p') \\ &+ f(p)\rho_1\sigma]v(p')e^{\frac{i}{h}[(E+h\nu-E')t-(p+n\frac{h\nu}{c}-p')r]} \\ &+ \text{complex conjugate part}\Big\}\,,\end{aligned} \tag{30}$$

where $\mathcal{J}_0$ denotes the current density belonging to the unperturbed eigenfunctions φ_0, ψ_0.

Now let $\mathcal{U}'$ be the vector potential belonging to $\mathcal{J}$ at a point whose distance r is very large compared to the dimensions of the region available for the electron, which in turn are to be regarded as large in comparison with the wavelength of the light and the de Broglie waves. These assumptions should just correspond to the observations of the Compton effect. Further, let n' be a unit vector which gives the direction of observation, i.e. the direction of a radius vector drawn from a point in the region of the electron to the point of observation. In a well-known way one then gets[7]

$$\begin{aligned}\mathcal{U}' = \frac{e}{r}\int\int dpdp'\Big\{&e^{\frac{i}{h}(E+h\nu-E')\left(t-\frac{r}{c}\right)}\int dru(p)[\rho_1\sigma g(p') \\ &- f(p)\rho_1\sigma]v(p')e^{-\frac{i}{h}\left[p-p'+n\frac{h\nu}{c}-n'\frac{E+h\nu-E'}{c}\right]r} \\ &+ \text{complex conjugate part}\Big\}\,,\end{aligned} \tag{31}$$

where $\int dv$ indicates the integration $\int dx_1 \int dx_2 \int dx_3$ over the whole region at the electron's disposal. Following Gordon[8] one can evaluate this integral with the help of the Fourier theorem by introducing, instead of p and p', certain new vectors β and β' with components P_1, P_2, P_3, P_1', P_2', P_3' through the relations

$$\mathcal{B} = p + n\frac{h\nu}{c} - n'\frac{E + h\nu}{c}, \qquad \mathcal{B}' = p' - \frac{E'}{c}n'. \tag{32}$$

It follows that

$$\mathcal{U}' = \frac{(2\pi h)^3}{r}\int \frac{d\mathcal{B}}{\Delta\Delta'}\Big\{e^{i\nu'\left(t-\frac{r}{c}\right)}u(p)[\rho_1\sigma g(p') + f(p)\rho_1\sigma]v(p') + \text{complex conjugate part}\Big\}, \tag{33}$$

where Δ and Δ', respectively, mean the functional determinants of P_k with respect to p_k and P_k' with respect to p_k', while p' is the special value of this quantity which for a given p follows from the relation $P = P'$, i.e. from

$$p + n\frac{h\nu}{c} = p' + n'\frac{h\nu'}{c}, \tag{34}$$

where ν' is given by the relation

$$E + h\nu = E' + h\nu'. \tag{35}$$

These are just the well-known Compton relations which assign a definite final state to the state p and the incident light quantum for a given direction of observation.

From (33) there now follows, according to Gordon,[9] the following expression for the radiation potential $\mathcal{U}(p, p')$ belonging to a transition from the state p to the state p':

$$\mathcal{U}(p, p') = \frac{(2\pi h)^3}{r}\frac{1}{\sqrt{\Delta\Delta'}}\Big\{e^{i\nu'\left(t-\frac{r}{c}\right)}u(p)[\rho_1\sigma g(p') + f(p)\rho_1\sigma]v(p') + \text{complex conjugate part}\Big\}, \tag{36}$$

where $u(p)$, $v(p)$ and $u(p')$, $v(p')$ from now on are to be regarded as normalized.

The radiation field belonging to a definite Compton effect is completely determined by the quantity $\mathcal{U}(p, p')$. From it the magnetic field strength $\mathcal{H}(p, p')$ follows through the relation

$$\mathcal{H}(p, p') = \operatorname{rot}\mathcal{U}(p, p'), \tag{37}$$

and from this in turn the electric field strength $\mathcal{E}(p, p')$ through the relation

$$\mathcal{E} = [\mathcal{H}(p, p'), n'] . \tag{38}$$

We now must obtain the quantity $\mathcal{U}(p, p')$ or, better, $\mathcal{H}(p, p')$ in a form suitable for a numerical calculation of the intensity and polarization of the scattered radiation. To this end, we have to insert in (36) the expressions for $f, \bar{f}, g, \bar{g}$ given by (25) and then use the calculational rules for the matrices ρ and σ given by Dirac, in order to obtain the result in terms of ordinary numbers. In this connection we will assume for simplicity that the initial state denoted by p corresponds to an electron at rest. We denote the corresponding values of $\mathcal{U}(p, p')$ and $\mathcal{H}(p, p')$ by $\mathcal{U}_0$ and $\mathcal{H}_0$. Thus,

$$p = 0 , \qquad E = mc^2 \tag{39}$$

should be valid.

With the help of Eqs. (5) for u and v we can reduce all matrices appearing in the expression for $\mathcal{U}_0$ to the unit matrix and the matrix vector σ. Based on the relations (2) it is then possible to give the result in a expression linear in the components of σ. Here the following calculational rule drawn from Dirac,[10] which is intimately connected with quaternion calculus, is of use, and easily proven with the help of (2):

$$(\mathcal{B}\sigma)(\mathcal{C}\sigma) = (\mathcal{B}\mathcal{C}) + i(\sigma, [\mathcal{B}\mathcal{C}]) , \tag{40}$$

as well as the one following from it,

$$\sigma(\sigma\mathcal{B}) = \mathcal{B} + i[\sigma\mathcal{B}] ,$$

where $\mathcal{B}$ and $\mathcal{C}$ stand for two arbitrary vectors, commuting with σ.

As follows from (5), in the quantities f and g there appears, besides the matrix σ, the matrix $\rho_1\sigma$. In (36) these matrices are multiplied by $\rho_1\sigma$. The resulting matrix is, according to (2) and (40), respectively, a linear combination of the unit matrix, ρ_1, $\rho_1\sigma$ and σ. This matrix then comes to stand between $u(p)$ and $v(p')$ and we can show that both $u(p)\rho_1 v(p')$ and $u(p)\rho_1\sigma v(p')$ can be reduced to the quantities $u(p)v(p')$ and $u(p)\sigma v(p')$. Indeed, it follows from

$$u(p)(1 + \rho_3) = 0 ,$$
$$\{E'/c + \rho_1(\sigma p') + \rho_3 mc\}v(p') ,$$

through multiplication by $\rho_1 v(p')$ and $u(p)\rho_1$, respectively, that

$$u(p)(\rho_1 + i\rho_2)v(p') = 0\,,$$
$$u(p)\{E'/c\rho_1 + (\sigma p') - i\rho_2 mc\}v(p') = 0\,,$$

or through eliminating $v(p)\rho_2 u(p')$,

$$u(p)\rho_1 v(p') = -\frac{u(p)(p'\sigma)v(p')}{E'/c + mc}\,. \tag{41}$$

In a similar way one finds that

$$u(p)\rho_1 \sigma u(p') = -\frac{u(p)\sigma(\sigma p')v(p')}{E'/c + mc}\,, \tag{42}$$

a relation which according to (40) can be reduced to an expression linear in σ.

We shall introduce the abbreviations

$$\left.\begin{array}{ll} u(p)\sigma v(p') = s\,, & u(p')\sigma v(p) = \bar{s}\,, \\ u(p)v(p') \;= d, & u(p')v(p) \;= \bar{d}\,, \end{array}\right\} \tag{43}$$

where s and $\bar{s}$ denote two vectors, complex conjugate to each other, while d and $\bar{d}$ are two complex conjugate scalars. With the help of the relations (41) and (42) we now obtain for the potential $\mathcal{U}_0$ an expression linear in s, $\bar{s}$ and d, $\bar{d}$. We shall not write this down here; rather, we content ourselves with the magnetic field strength to be calculated from (27), which is decisive for both the polarization and the intensity of the scattered radiation. It follows after some calculation, where the relations (34) and (35) are employed,

$$\begin{aligned} \mathcal{H}_0 = {} & \frac{(2\pi h)^3 e^2 \nu'}{2mc^2 r\left(\nu - \nu' + \dfrac{2mc^2}{h}\right)} \sqrt{\frac{E'\nu'}{mc^2\nu}} \Bigg\{ d\left(\frac{1}{\nu}(n'\varepsilon)(\nu' - \nu)[n'n] \right. \\ & \left. - \nu'\left(\frac{1}{\nu} + \frac{1}{\nu'}\right)^2 \frac{mc^2}{h}[n'\varepsilon]\right) - i\Bigg[\left(\frac{1}{\nu'} - \frac{1}{\nu}\right)\Bigg((s,\, n\nu - n'\nu')((n'\varepsilon)n \\ & - (nn')\varepsilon) + \left(\nu - \nu' + \frac{2mc^2}{h}\right)((sn')\varepsilon - (n'\varepsilon)s)\Bigg) \\ & + \frac{2}{\nu}(\varepsilon n')((n's)(n\nu - n'\nu') + (\nu' - (nn')\nu)s) \\ & - \left(\frac{1}{\nu} + \frac{1}{\nu'}\right)\Big((n[\varepsilon s])\nu[n'n] + \nu'(n'[n\varepsilon])[n's]\Big)\Bigg]\Bigg\} e^{i\nu'\left(t - \frac{r}{c}\right)} \\ & + \text{complex conjugate part}\,, \end{aligned} \tag{44}$$

where we have used[11] the values for Δ and Δ' following from (32) with the help of (34), namely

$$\Delta = 1 - \frac{c}{E}(n'p) = 1, \quad \Delta' = 1 - \frac{c}{E'}(n'p') = \frac{mc^2 v}{E'v'} . \tag{45}$$

We will not here enter into a closer discussion of this expression, but rather content ourselves with investigating the intensity of the scattered radiation in various directions.[12] To this end we form the quantity $\mathcal{H}_0^2$. In this context we want to choose as initial state one of the eigenfunctions belonging to $p = 0$ mentioned on page 4, i.e. for this state either only u_1, v_1 or only u_2, v_2 are different from zero and normalized according to (20). Thus, we look at the scattered radiation from an electron which has been magnetized by a field in the direction of the x_3-axis. For the final state we have to assume here that both u_1^*, v_1^* and u_2^*, v_2^* are different from zero and normalized. According to the usual quantum-mechanical methods for calculating transition probabilities, this corresponds to the possibility of such transitions where the magnetic moment of the electron is unchanged as well as such where its direction is changed.

By forming $\mathcal{H}_0^2$ we now obtain an expression bilinear in the quantities d, s and $\bar{d}$, $\bar{s}$, which we have to average over the phases δ_1 and δ_2 of the initial and final states. Thus we are dealing with a sum of terms of the form $u(p')\alpha v(p).u(p)\beta v(p')$ which we have to average with respect to the phases, where α and β are two matrices. If we first assume that $u_1(p)$ and $v_1(p)$ are different from zero, we write this as

$$\sum_{i,k} u_i(p')\alpha_{i1} v_1(p) u_1(p) \beta_{1k} v_k(p') .$$

This, however, is equal to $u_1(p)v_1(p) \cdot u(p')\alpha\mu\beta v(p')$, where μ represents the matrix

$$\begin{pmatrix} 1 & 0 & 0 & 0 \\ 0 & 0 & 0 & 0 \\ 0 & 0 & 0 & 0 \\ 0 & 0 & 0 & 0 \end{pmatrix} .$$

We can, according to (8) and (9), represent μ in the following way with the help of ρ_3 and σ_3:

$$\mu = \left(\frac{1+\sigma_3}{2}\right)\left(\frac{1-\rho_3}{2}\right) , \tag{46}$$

and it thus holds that

$$u(p')\alpha v(p) \cdot u(p)\beta v(p') = (2\pi h)^{-3} u(p')\alpha\mu\beta v(p'), \tag{47}$$

where we, according to (20), have put $u_1 v_1 = (2\pi h)^{-3}$. Had we chosen as initial state the state belonging to u_2, v_2, the quantity to take for μ would have been $(\frac{1-\sigma_3}{2})(\frac{1-\rho_3}{2})$.

We now assume that $\alpha = (\mathcal{C}\sigma)$ and $\beta = (\mathcal{B}\sigma)$, where $\mathcal{B}$ and $\mathcal{C}$ are vectors commuting with σ. Then $u(p')\alpha v(p) = (\mathcal{C}\bar{s})$ and $u(p)\beta v(p') = (\mathcal{B}s)$, and we have

$$(\mathcal{B}s)(\mathcal{C}\bar{s}) = (2\pi h)^{-3} u(p')(\mathcal{C}\sigma)\mu(\mathcal{B}\sigma)v(p').$$

Now the relation

$$(\mathcal{C}\sigma)\left(\frac{1+\sigma_3}{2}\right) = \left(\frac{1-\sigma_3}{2}\right)(\mathcal{C}\sigma) + C_3$$

holds, so that

$$(\mathcal{B}s)(\mathcal{C}\bar{s}) = (2\pi h)^{-3} u(p')\left(\frac{1-\rho_3}{2}\right)\left\{\left(\frac{1-\sigma_3}{2}\right)(\mathcal{C}\sigma)(\mathcal{B}\sigma) + C_s(\mathcal{B}\sigma)\right\}v(p'). \tag{48}$$

On the right-hand side of this equation there stand between $u(p')$ and $v(p')$ matrices, which are either of the form $\frac{1-\rho_3}{2}$ or of the form $\frac{1-\rho_3}{2}\sigma$. We can easily transform these in the following way. From the equations

$$u(p')\{E'/c + \rho_1(\sigma p') + \rho_3 mc\} = 0, \quad \{E'/c + \rho_1(\sigma p') + \rho_3 mc\}v(p') = 0$$

it follows, in the same way as on page 11, that

$$u(p')\left(\frac{1-\rho_3}{2}\right)v(p') = \frac{1}{2}\left(1 + \frac{mc^2}{E'}\right)u(p')v(p'). \tag{49}$$

Further, it follows that

$$u(p')\left(\frac{1-\rho_3}{2}\right)\sigma v(p') = \frac{1}{2}\left(1 + \frac{E'}{mc^2}\right)u(p')\sigma v(p') - \frac{p'}{2mE'}u(p')(\sigma p')v(p'), \tag{50}$$

so that the quantities $u(p')(\frac{1-\rho_3}{2})v(p')$ and $u(p')(\frac{1-\rho_3}{2})\sigma v(p')$ can be expressed through the quantities $u(p')v(p')$ and $u(p')\sigma v(p')$. As now the final state characterized by p' should contain the two independent solutions in equal strength, it is clear from the outset that the average of $u(p')\sigma v(p')$ over the phases must vanish. In fact, this also follows from the representations (18) and (21) of the quantities $u(p')$, $v(p')$. Thus, we assume that

$$\overline{u(p')\sigma v(p')} = 0\,, \tag{51}$$

where the bar indicates the averaging over the phases $\delta_1(p')$ and $\delta_2(p')$. This assumption is now enough to enable the calculation of all other averages appearing. First we must have the value of $u(p')v(p')$. It follows from (18) and (20) that

$$u(p')v(p') = u^*(p')v^*(p') = 2(2\pi h)^{-3}\,. \tag{52}$$

Further, it follows from (48) that

$$\overline{(\mathcal{B}s)(\mathcal{C}\bar{s})} = \frac{1}{2}(2\pi h)^{-3}u(p')\left(\frac{1-\rho_3}{2}\right)v(p')\{(\mathcal{B}\mathcal{C}) + i[\mathcal{B}\mathcal{C}]_3\}\,, \tag{53}$$

or, according to (49) and (52), if we put

$$\gamma = \frac{1}{2}(2\pi h)^{-6}\left(1 + \frac{mc^2}{E'}\right), \tag{54}$$

$$\overline{(\mathcal{B}s)(\mathcal{C}\bar{s})} = \gamma\{(\mathcal{B}\mathcal{C}) + i[\mathcal{B}\mathcal{C}]_3\}\,. \tag{55}$$

We can easily liberate ourselves from the special choice of coordinates. If $\mathcal{I}$ is a unit vector in the direction of the magnetic moment of the magnetized electron in the initial state, we obviously have in general

$$\left.\begin{aligned} &\overline{(\mathcal{B}s)(\mathcal{C}\bar{s})} = \gamma\{(\mathcal{B}\mathcal{C}) + i(\mathcal{I}[\mathcal{B}\mathcal{C}])\}\,, \\ \text{and similarly}\quad &\overline{[\mathcal{B}s][\mathcal{C}\bar{s}]} = \gamma\{2(\mathcal{B}\mathcal{C}) + i(\mathcal{I}[\mathcal{B}\mathcal{C}])\}\,, \\ \text{also}\quad &\overline{(\mathcal{B}s)\bar{d}} = \overline{d(\mathcal{B}\bar{s})} = \gamma(\mathcal{I}\mathcal{B})\,, \end{aligned}\right\} \tag{56}$$

and finally

$$\overline{d\bar{d}} = \gamma\,. \tag{57}$$

When calculating $\overline{\mathcal{H}_0^2}$ we now assume for simplicity that the incident light linearly polarized so that $\epsilon = \bar{\epsilon}$. It then follows after some calculation that

$$\overline{\mathcal{H}_0^2} = \frac{e^4}{m^2c^4r^2}\left(\frac{\nu'}{\nu}\right)^3\left\{\left(\frac{\nu}{\nu'} + \frac{\nu'}{\nu}\right)\varepsilon^2 - 2(n'\varepsilon)^2\right\} \tag{58}$$

or, if we denote the angle between the direction of observation and the wave normal of the incident light by Θ and the angle between the direction of observation and the electric force of the incident wave by ϑ,

$$I = I_0 \frac{e^4}{m^2c^4r^2} \frac{\sin^2\vartheta}{(1+\alpha(1-\cos\Theta))^3} \left(1+\alpha^2 \frac{(1-\cos\Theta)^2}{2\sin^2\vartheta(1+\alpha(1-\cos\Theta))}\right), \quad (59)$$

where $\alpha = \frac{h\nu}{mc^2}$, while I_0 denotes the intensity of the incident radiation, and I the intensity of the scattered radiation, of which the former stands in the same relation to $2\epsilon^2$ as the latter to $\overline{\mathcal{H}_0^2}$. If the incident radiation is unpolarized we must further average over all directions of ϵ for a given direction of the incident radiation. The average of $\sin^2\vartheta$ here becomes $\frac{1}{2}(1+\cos^2\Theta)$, so that

$$\bar{I} = I_0 \frac{e^4}{2m^2c^4r^2} \frac{1+\cos^2\Theta}{(1+\alpha(1-\cos\Theta))^3} \left(1+\alpha^2 \frac{(1-\cos\Theta)^2}{(1+\cos^2\Theta)(1+\alpha(1-\cos\Theta))}\right). \quad (60)$$

If we look at the last formulae, (58)–(60), we see that the vector $\mathcal{I}$ does not appear therein. The results are thus independent of whether the scattering electrons have first become magnetized or not. As is gathered from a closer look at the expression (44), this is, however, valid in general only if the primary radiation is linearly polarized. The scattering for elliptically polarized incoming light is hence not equal to the sum of the scatterings belonging to the individual polarized components, a circumstance which is intimately connected with the curious polarization of the scattered radiation.[13]

Further, we see that the expressions (59) and (60) only differ through the factors

$$\left(1+\alpha^2 \frac{(1-\cos\Theta)^2}{2\sin^2\vartheta(1+\alpha(1-\cos\Theta))}\right)$$

and

$$\left(1+\alpha^2 \frac{(1-\cos\Theta)^2}{(1+\cos^2\Theta)(1+\alpha(1-\cos\Theta))}\right),$$

respectively, from the corresponding formulas of the Dirac–Gordon theory, i.e. the deviations of the two theories are of the order of magnitude of $\left(\frac{h\nu}{mc^2}\right)^2$, while the deviations from the earlier expressions of the classical theory of scattered radiation, developed by J. J. Thomson, are of the order of magnitude of[14] $\frac{h\nu}{mc^2}$.

From (60) we can easily derive an expression for the scattering coefficient of a substance which contains N electrons per unit volume. Namely, if we multiply the expression (60) by $\frac{d\Omega}{h\nu'}$, we obtain the number of quanta scattered by an electron in a solid angle $d\Omega$ in the direction Θ. To each scattered light quantum belongs, however, an incoming light quantum of the magnitude $h\nu$, so that the energy loss of the incident radiation through scattering in this direction on an electron is equal to $\frac{\nu}{\nu'}d\Omega$ times the expression (60). By integrating over all directions and multiplying by the number of electrons per unit volume, we thus obtain for the scattering coefficient

$$S = \frac{2\pi N e^4}{m^2 c^4}\left\{\frac{1+\alpha}{\alpha^2}\left[\frac{2(1+\alpha)}{1+2\alpha} - \frac{1}{\alpha}\log(1+2\alpha)\right] + \frac{1}{2\alpha}\log(1+2\alpha) - \frac{1+3\alpha}{(1+2\alpha)^2}\right\}, \qquad (61)$$

which again differs from the corresponding expression given by Dirac by quantities of order α^2.

Copenhagen, Universitetets Institut for teoretisk Fysik, October 1928

References

1. P. A. M. Dirac, *Proc. R. Soc.* **A111**, 405 (1926); will be quoted as A in the following.
2. W. Gordon, *Z. Phys.* **40**, 117 (1927).
3. P. A. M. Dirac, *Proc. R. Soc.* **A117**, 610 (1928); will be quoted as B in the following.
4. Y. Nishina, *Z. Phys.* **52**, 869 (1929).
5. P. A. M. Dirac, *loc. cit.* B.
6. See: O. Klein, *Z. Phys.* **41**, 407, 1927.
7. See e.g.: O. Klein, therein, *loc. cit.*, p. 422.
8. W. Gordon, *loc. cit.*, p. 129.
9. W. Gordon, *loc. cit.*, p. 130; see also: I. Waller, *Phil. Mag.* **4**, 1228, 1927.
10. P. A. M. Dirac, *loc. cit.* B., p. 618.
11. W. Gordon, *loc. cit.*, p. 130.
12. Cf. the accompanying work by Y. Nishina, where the question of the polarization of the scattered radiation is discussed.
13. Cf. Y. Nishina, *loc. cit.*, where this point has been more closely investigated.
14. Cf. *Nature* **122**, 398 (1928), where the relationship between our results and experiment is briefly treated.

THE REFLECTION OF ELECTRONS AT A POTENTIAL JUMP ACCORDING TO DIRAC'S RELATIVISTIC DYNAMICS*

O. KLEIN

in Copenhagen

Received December 24, 1928

The reflection of electrons at a potential jump is investigated according to the new dynamical theory of Dirac. At very large values of the potential jump, electrons may according to the theory penetrate against the electric force acting on them through the surface of the jump and arrive at the other side with a negative kinetic energy. This could be regarded as a particularly sharp example of the difficulty emphasized by Dirac of the relativistic dynamics.

1. Introduction

As Dirac[1] has emphasized, a serious difficulty of the relativistic quantum theory consists in the circumstance that an electron in a field of force may, according to the theory, acquire negative values of the energy, which are in general connected with the physically meaningful positive energy values through the possibility of transitions. Also in his new, in other respects so successful treatment of relativistic quantum dynamics, he has not succeeded in overcoming this difficulty. In the following lines an elementary example will be pointed out where this difficulty appears in a particularly sharp way. This has to do with the reflection and refraction of electron waves at a boundary surface, where the electrostatic potential has a jump.

2. Let E be the total energy of an electron moving in a force-free part of space, while p_1, p_2, p_3 may denote the components of its momentum with respect to the axes of an orthogonal coordinate system, where the electron has

*Original in *Z. Phys.* **53**, 157 (1929); reproduced here with permission from Springer-Verlag. Translated from the German by Dr Lars Bergström.

the coordinates x_1, x_2, x_3. We will assume that the electrostatic potential is different from zero in this part of space; in fact the electron should have the constant potential energy P. This specification of course only has a meaning if we compare this part of space with another part of space, where the potential has another value. According to the ordinary relativistic mechanics the following relationship holds between the energy $E - P$, which we will call the kinetic energy of the electron (although for an electron at rest it is not zero, but rather m_0c^2), and the momentum

$$\left(\frac{E-P}{c}\right)^2 = p_1^2 + p_2^2 + p_3^2 + m_0^2c^2 \,, \tag{1}$$

where m_0 stands for the rest mass of the electron and c the velocity of light. The difficulty in question is connected to the fact that the kinetic energy can take both positive and negative values, as a result of which, besides the physically meaningful solutions, further solutions are present, which cannot be assigned a physical meaning. In the usual relativistic mechanics there is no difficulty with this as the square of the momentum can never become negative so that according to (1) the kinetic energy can never become zero; since in this theory only continuous transitions occur, this means that the negative values of energy can never be reached. In the quantum theory, however, the solutions in question cannot in general be separated, since on the one hand, discontinuous transitions are possible and, on the other hand, the electron waves here can penetrate through regions where the electron, classically speaking, has an imaginary momentum.

3. For an electron in an electrostatic force field, where the potential is V, we can according to Dirac reduce the problem to the following wave equation:

$$\left\{\frac{E+eV}{c} + \beta mc\right\}\psi - ih\sum_1^3 \alpha_k \frac{\partial\psi}{\partial x_k} = 0 \tag{2}$$

with the adjoint equation

$$\varphi\left\{\frac{E+eV}{c} + \beta mc\right\} + ih\sum_1^3 \frac{\partial\varphi}{\partial x_k}\alpha_k = 0 \,, \tag{2a}$$

where again E means the total energy of the electron, which we view as being given, while $-e$ denotes its charge and h is the Planck constant divided by

2π. The quantities α_1, α_2, α_3 and β are matrices with four rows and columns, fulfilling the relations

$$\alpha_i\alpha_k + \alpha_k\alpha_i = 0, \quad i \neq k, \quad \alpha_i\beta + \beta\alpha_i = 0\,,$$
$$\alpha_1^2 = \alpha_2^2 = \alpha_3^2 = \beta^2 = 1\,. \tag{3}$$

Correspondingly, the functions φ and ψ consist of four components each, $\varphi_1, \varphi_2, \varphi_3, \varphi_4$ and $\psi_1, \psi_2, \psi_3, \psi_4$, respectively. With γ denoting a matrix with four rows and columns, $\gamma\psi$ shall here stand as an abbreviation for the four quantities $\sum_{k=1}^{4} \gamma_{ik}\psi_k$ $(i = 1, 2, 3, 4)$, where γ_{ik} denotes the matrix elements of γ. In the same way, $\varphi\gamma$ should be interpreted as $\sum_{k=1}^{4} \varphi_k\gamma_{ki}$ $(i = 1, 2, 3, 4)$. One sees that it is then allowed to multiply (2) from the left and (2a) from the right by any matrix without spoiling their validity. This just means a linear transformation of the system of equations.

We now want to assume that to the left of the plane $x_1 = 0$ the potential V equals zero, whereas right of this plane $eV = -P$ is valid, where P means a positive quantity. One would thus expect that the electrons lose a part P of their kinetic energy when passing this plane. To be able to investigate the reflection and refraction of electron waves at this surface of discontinuity, it is necessary to find the matching conditions of the Dirac wave equations at such surfaces. One can, as is customary,[2] derive this by regarding the surface of discontinuity as a limiting case of a region of finite thickness in which the discontinuous quantity, in this case the potential, changes rapidly. Since Eqs. (2) can be solved with respect to the derivatives of the components of ψ perpendicular to the surface of discontinuity, in this case $\frac{\partial\psi_1}{\partial x_1}, \frac{\partial\psi_2}{\partial x_1}, \frac{\partial\psi_3}{\partial x_1}, \frac{\partial\psi_4}{\partial x_1}$, it follows immediately, if only the potential remains finite in the transition region, that the four quantities $\psi_1, \psi_2, \psi_3, \psi_4$ and of course also $\varphi_1, \varphi_2, \varphi_3, \varphi_4$ remain continuous at the passage of the surface of discontinuity.[a]

4. Without any important restriction we can now look at a purely harmonic incident wave which hits the plane $x_1 = 0$ perpendicularly. We thus set for the incident wave, where we simply write x for x_1,

$$\psi_e = v_e e^{\frac{i}{\hbar}(px - Et)}\,, \tag{4}$$

[a]Through the solubility of Eq. (2) as regards the derivatives of the four components of ψ from an arbitrary direction in space, it also follows that for a continuous solution not all four components can equal zero at a surface without vanishing completely. The boundary condition $\psi = 0$ used for the Schrödinger equation at a wall is thus senseless in the theory of Dirac and must be replaced by conditions which are to be searched for in a more complete specification of the physical properties of the wall.

where t denotes the time and p the momentum of the electron. By inserting this expression in (2) we obtain a system of linear algebraic equations for the four components of the amplitude v_e, which can be written in the following way:

$$\{E/c + \alpha p + \beta m_0 c\} v_e = 0\,, \tag{5}$$

where we have put α for α_1. If v_e should not vanish identically, the condition below follows:

$$E^2/c^2 = p^2 + m_0^2 c^2\,, \tag{6}$$

which is a special case of (1). We will here choose for E the positive value. Now one can freely choose two of the components of ψ, which just corresponds to the two possibilities of orientation of the electron in a magnetic field.[3] From (6) it follows that the momentum of the reflected wave has to be $-p$, while for the refracted wave a value of momentum $\bar{p}$ follows, given by the relationship

$$\left(\frac{E-P}{c}\right)^2 = \bar{p}^2 + m_0^2 c^2\,. \tag{7}$$

We first want to assume P to be so small that a positive value of $\bar{p}^2$ follows from (7). We can then put

$$\psi_r = v_r e^{\frac{i}{\hbar}(-px - Et)}, \quad \psi_g = v_g e^{\frac{i}{\hbar}(\bar{p}x - Et)}\,, \tag{8}$$

where ψ_r and ψ_g are associated with the reflected and refracted wave, respectively. From (2) follows

$$\left\{\frac{E}{c} - \alpha p + \beta m_0 c\right\} v_r = 0, \quad \left\{\frac{E-P}{c} + \alpha \bar{p} + \beta m_0 c\right\} v_g = 0\,. \tag{9}$$

The boundary condition now simply reads

$$v_e + v_r = v_g\,. \tag{10}$$

If we regard the four components of the incident wave as given, we have eight unknowns, namely the four components of v_r and the four components of v_g. However, because of (9) only four of these are independent so that (10) gives just the sufficient number of equations for their calculation. We can easily obtain the solution of the equations with respect to v_r in the following way. From (5) and the first equation of (9) follows

$$(E/c + \beta m_0 c)(v_e + v_r) = -\alpha p (v_e - v_r)\,.$$

From (9) it follows, based on (10), that

$$(E/c + \beta m_0 c)(v_e + v_r) = (P/c - \alpha\bar{p})(v_e + v_r)$$

and thus

$$(P/c - \alpha\bar{p})(v_e + v_r) = -\alpha p(v_e - v_r)$$

or

$$\{P/c - \alpha(p + \bar{p})\}v_r = -\{P/c + \alpha(p - \bar{p})\}v_e\,.$$

Through multiplication of both sides of this relation by $\frac{P}{c} + \alpha(p+\bar{p})$, it follows that, by taking into account $\alpha^2 = 1$ with the help of (6) and (7),

$$v_r = -\frac{2P/c(E/c + \alpha p)}{P^2/c^2 - (p+\bar{p})^2}\, v_e\,. \tag{11}$$

To be able to utilize this result physically we must seek the corresponding solution to the adjoint wave equation (2a), since according to Dirac $\varphi\psi dv = \sum_1^4 \varphi_k\psi_k dv$ gives the probability that we find the electron in the volume element dv. From this follows[4] for the probability that the electron traverses a surface element df perpendicular to the x-axis in the time dt, $-c\varphi\alpha\psi df\,dt$, where $\varphi\alpha\psi$ is an abbreviation for $\sum_{i,k=1}^4 \varphi_i\alpha_{ik}\psi_k$. As Dirac has shown, it is now possible to choose Hermitian matrices for α and β. If ψ stands for a solution of (2), then $\varphi =$ complex conjugate of ψ will be a solution of (2a). As we choose with Hermitian matrices complex conjugate values for φ and ψ, we obviously get real expressions for $\varphi\psi$ and $\varphi\alpha\psi$. We put accordingly

$$\varphi_e = u_e e^{-\frac{i}{\hbar}(px - Et)}, \quad \varphi_r = u_r e^{-\frac{i}{\hbar}(-px - Et)},$$
$$\varphi_g = u_g e^{-\frac{i}{\hbar}(\bar{p}x - Et)}, \tag{12}$$

where, if α and β are Hermitian, u_e, u_r and u_g are the complex conjugates of v_e, v_r and v_g. From (2a) and (12) follows

$$u_e\{E/c + \beta m_0 c - \alpha p\} = 0, \quad u_r\{E/c + \beta m_0 c + \alpha p\} = 0,$$
$$u_g\left\{\frac{E - P}{c} + \beta m_0 c + \alpha\bar{p}\right\} = 0\,. \tag{13}$$

We now derive from (5) and (13) a useful identity as we multiply (5) from the left by $u_e\alpha$ and the first equation (13) from the right by αv_e. Since α and β anticommute and $\alpha^2 = 1$, we get by addition

$$E/c\, u_e\alpha v_e + p u_e v_e = 0$$

or

$$-cu_e\alpha v_e = \frac{pc^2}{E} u_e v_e \,. \tag{14}$$

In the particle interpretation $\frac{pc^2}{E}$ means the velocity of the electron (group velocity), so that this equation gives a connection between current density and density in correspondence with the usual hydrodynamics. A similar result is of course obtained for the reflected and refracted waves (for the latter E is to be replaced by the kinetic energy $E - P$). To calculate the fraction of electrons that is reflected and refracted, respectively, it is thus enough to express the quantities $u_r v_r$ and $u_g v_g$ in terms of the components of the incoming wave.

Through a similar calculation to that which led to the expression (11) we now find

$$u_r = -u_e \frac{2P/c(E/c + \alpha p)}{P^2/c^2 - (p+\bar{p})^2} \,. \tag{15}$$

From (10) and (15) then follows

$$\begin{aligned} u_r v_r & \left(\frac{2P/c}{P^2/c^2 - (p+\bar{p})^2}\right)^2 u_e(E/c + \alpha p)^2 v_e \\ &= \left(\frac{2P/c}{P^2/c^2 - (p+\bar{p})^2}\right)^2 \left\{(E^2/c^2 + p^2)u_e v_e + \frac{2Ep}{c} u_e \alpha v_e\right\} \end{aligned}$$

or according to (14) and (6)

$$u_r v_r = \left(\frac{2Pm_0}{P^2/c^2 - (p+\bar{p})^2}\right)^2 u_e v_e \,. \tag{16}$$

The quantity $\left(\frac{2Pm_0}{P^2/c^2-(p+\bar{p})^2}\right)^2$ thus gives the fraction of electrons that becomes reflected. As one can easily prove with the help of (6) and (7), the reflection coefficient grows with increasing P from zero for $P = 0$ and reaches the value of one for $P = E - m_0c^2$. Just here $\bar{p}^2$ now equals zero, and by further increase of P we enter into the region of imaginary $\bar{p}$, which we shall now investigate.

In the classical theory, the fact that $\bar{p}$ becomes imaginary means that the electron penetrates so deep into the field that its velocity becomes zero and it is then thrown back. In the wave theory the wave functions will have finite values also to the right of the boundary surface; as we shall see, however, the conditions primarily correspond to total reflection in optics.

If $\bar{p}$ is imaginary, we can put

$$\psi_g = v_g e^{-\mu x - i\frac{E}{h}t}, \; \varphi_g = u_g e^{-\mu x + i\frac{E}{h}t}, \tag{17}$$

where μ means a real quantity which must obviously be positive as otherwise the density to the right of the boundary surface would grow to infinity with x. Just because in this case μ is real, the exponent proportional to x in ψ_g and φ_g must according to the general prescriptions of the theory have the same sign. This means, however, that we put for ψ_g the quantity $\bar{p}$ equal to $ih\mu$, for φ_g on the other hand equal to $-ih\mu$. With this prescription we obtain from (11) and (15)

$$v_r = -\frac{2P/c(E/c+\alpha p)v_e}{P^2/c^2-(p+ih\mu)^2}, \quad u_r = -u_e\frac{2P/c(E/c+\alpha p)}{P^2/c^2-(p-ih\mu)^2} \tag{18}$$

and thus

$$u_r v_r = \frac{(2P/c)^2(E^2/c^2-p^2)}{[(P/c+p)^2+\mu^2h^2][(P/c-p)^2+\mu^2h^2]}u_e v_e\,.$$

We can simplify this expression with the help of (6) and (7). First follows

$$\bar{p}^2 = p^2 - \frac{P(2E-P)}{c^2} \tag{19}$$

and thus with $\bar{p}^2 = -\mu^2h^2$

$$(P/c\pm p)^2+\mu^2h^2 = 2P/c(E/c\pm p)\,.$$

Therefore it simply follows that

$$u_r v_r = u_e v_e\,. \tag{20}$$

Thus the reflected current equals the incident current, while an exponentially decreasing wave solution exists behind the boundary surface. The condition for this case is according to (19) $p^2 < \frac{P(2E-P)}{c^2}$, a condition which with increasing P is first fulfilled when P exceeds the value $E - c\sqrt{E^2/c^2-p^2} = E - m_0c^2$. If P grows even further the quantity μ will first increase; due to the quadratic term in P in (19) it reaches however a maximum, which occurs for $P = E$. From there on μ becomes smaller and is again zero for $P = E + c\sqrt{E^2/c^2-p^2} = E + m_0c^2$. For even larger P, $\bar{p}$ again takes real values so that Eqs. (11), (15) and (16) again constitute the solution of the problem. In this region the kinetic energy $E-P$ is instead negative, so that we really have reached the mechanically forbidden region. This has the consequence that the group velocity, which is given by $\frac{c^2}{E-P}\bar{p}$, is directed opposite to the momentum and we must take[b] a negative value for $\bar{p}$. This one can easily see if one takes as initial state a wave packet which moves towards the boundary surface from the left.

We have thus arrived at the curious result that for values of P larger than $E+m_0c^2$, a fraction of the electrons crosses the potential threshold, while their kinetic energy is transformed from the original positive to a negative value. It is of interest to calculate the group velocity of these through-going electrons. For these it follows from (7) that

$$\frac{c^2}{P-E}|\bar{p}| = c\sqrt{1-\left(\frac{m_0c^2}{P-E}\right)^2}\,. \tag{21}$$

For $P = E + m_0c^2$ this velocity is, as could be expected, just equal to zero. It then grows with growing P to reach the velocity of light for $P = \infty$.

As one can see from the expression (16), the coefficient of reflection, which for $P = E + m_0c^2$ equals unity, decreases gradually for increasing P, to the value $\frac{E/c-p}{E/c+p}$ for $P = \infty$. The corresponding limiting value for the fraction of electrons that penetrates the boundary surface is thus $\frac{2p}{E/c+p}$, i.e. of the same order of magnitude as the ratio between the velocity of the incident electrons and the velocity of light, and can, for large values of p, attain considerable values. For $p = m_0c$, corresponding to a velocity of the incident electrons of around 70% of the velocity of light, we get for instance the value $2(\sqrt{2}-1)$, i.e. around 83%. It is of course not important that we have here assumed $P = \infty$; evidently one would get numbers of the same order of magnitude as soon as P is several times larger than the rest mass m_0c^2 of the electrons. We will not here enter into the question whether it is possible to realize such potential jumps experimentally. It should only be emphasized that the difficulty in question is not connected to the assumption of a discontinuity, which was chosen only for mathematical simplicity. Also if the jump surface is replaced by a small region where the potential grows rapidly but continuously, electrons will, as follows from this whole procedure of calculation, penetrate into the forbidden region where they possess negative kinetic energy. This is closely connected to the fact that in the case discussed above the total reflection of the wave solution behind the boundary surface does not disappear even though the electron here would have an imaginary momentum according to classical mechanics. It should also be mentioned that following from the expression (16)

[b]I had originally not taken this circumstance into account; rather, it arose in a conversation with Mr W. Pauli, whom I here would like to cordially thank.

the reflection coefficient of electrons that fall towards the boundary surface with momentum $\bar{p}$ from a part of space where the potential energy is P, has the same value of the reflection coefficient for the reversed process.

As a result of our investigations we can thus establish that the difficulty of the relativistic quantum mechanics emphasized by Dirac can appear already in purely mechanical problems where no radiation processes are involved.

At the end of this note I would like to cordially thank Professor Niels Bohr for many conversations, which have contributed significantly towards clarifying the considerations above.

Copenhagen, Universitetets Institut for teoretisk Fysik, December 1928

References

1. P. A. M. Dirac, *Proc. R. Soc.* **117**, 612 (1928).
2. Cf. H. Faxén and J. Holtsmark, *Z. Phys.* **45**, 311 (1927), where a similar study was performed for the wave equation of Schrödinger.
3. Cf. C. G. Darwin, *Proc. R. Soc.* **118**, 654 (1928).
4. P. A. M. Dirac, *Proc. R. Soc.* **118**, 351 (1928).

EXCERPTS FROM SOME AUTOBIOGRAPHICAL NOTES*

O. KLEIN
in Stockholm

Not until the summer of 1935 did I make new attempts concerning the five-dimensional theory, in connection with a conversation with Pauli, who had in the meantime himself published two articles containing a modified form thereof, which had been proposed by Veblen. However, it was not until 1937 that I once again took hold of it — this time in connection with Yukawa's idea concerning the interaction of the nucleons through an intermediate particle, which gradually appeared to be essentially correct and utterly fertile for all of nuclear physics.

As concerns this work, about which I gave a talk in Warsaw in the summer of 1938 at a meeting on the new theories of physics (published in *Institut International de Coopération Intellectuelle, Collection Scientifique*, 1939),[1] I have to say that this — although it pointed towards aspects that later, in another context, have become of importance — was suffering from the fact that I, like on some other occasions, tried to apply a mathematical formalism to the too incomplete knowledge one had at the time of what is now called elementary particle physics. However, for me it became of use as a background for much later work.

These later works, like the one mentioned, have as a starting point Dirac's electron theory in a general relativistic form and at the same time my old five-dimensional studies, which I therefore have expanded upon very thoroughly in the preceding work.[2] In their present form we are first of all dealing with an extension of the principle of equivalence of Einstein, fundamental for gravita-

*Original in *Svensk Naturvetenskop*, 1973, reproduced here with permission of the Swedish Science Research Council. Translated from the Swedish by Dr Lars Bergström.

tion. My further occupation with this principle was, however, connected to another area of problems, named relativistic cosmology.

During earlier years I had written a small book on Einstein's theory of relativity (*Natur och Kultur*, 1933), in which I — due to more general doubts — had excluded Einstein's cosmological attempts, which constituted the introduction to the more recent cosmology. On the other hand, I had thoroughly accounted for Mach's idea about the inertial forces as a kind of gravitational action from the assembly of distant celestial bodies in the Universe, which was the main background for this attempt.

That later I was to occupy myself in detail with these questions was due to the following: around the mid-1940's my collaborators (Göran Beskow and Lars Treffenberg) and I had further developed an attempt to explain the relative abundance of the elements as a result of an early thermal equilibrium at very high temperature in a kind of primordial stars. The result looked promising, but simultaneously difficulties appeared. At the same time Gamow and his collaborators Alpher and Herman had tried to explain the same thing in a completely different way: through a series of nuclear reactions during a very early epoch of the evolution of the Universe according to the expanding relativistic cosmology, usually called the Big Bang. After an encounter with them in America in the beginning of 1950, it was clear to me that their attempt was carried out very well, that the results were about as promising as ours, but that they too had encountered difficulties.

This equivalence, as it appeared to be for better or worse, between two so different attempts, made me start becoming more interested in the question of the state of the world at the time of formation of the elements. Thus, I had started to become interested in cosmology, but was still sceptical due to the great difference between it and ordinary physics ever since the days of Galileo — where the concern was about finding ever more general laws of nature and not the structure of the whole world. In other words, cosmology appeared unnatural to me.

Already somewhat earlier I had convinced myself that a system of stars, formed in a natural way through the contraction of a cloud of small density, cannot be compressed indefinitely. The average density for a given mass has a particular limit which is inversely proportional to the square of the mass. For a system of the size of the Milky Way the average density can thus not be greater than about one part in 100 000 of the density of water. Now it seems that the part of the Universe reached by telescopes contains around ten billion Milky Ways, which would mean that its average density before the start of the expansion can hardly have been greater than 10^{-25} g/cm^3, a remarkably good vacuum! The most important conclusions from the Big Bang cosmology, an eventual formation of helium and above all the so-called fireball radiation, which following a suggestion by Alpher and Herman should now have cooled to a temperature a few degrees above absolute zero — and which after the observation of such a radiation constitutes the main argument of cosmologists — is, according to the theory, connected to a state, the density of which is several orders of magnitude higher than that of water.

In the meantime I had, however, found still another reason to replace the infinite Universe of cosmology with a model for the expanding system of Milky Ways — a meta-galaxy — which certainly is very large yet finite, such that it like all other systems of stars has originated in a natural way. Probably there exist also other such Milky Way systems, although they have not been observed so far. My starting point for this consisted of some relations of orders of magnitude found by the great astronomer Eddington, according to which there would be a relation between the mass and dimensions of a closed Universe on the one hand, and similar atomic quantities on the other, that were regarded by him as a deep connection between micro-cosmos and macro-cosmos. In both these relations there appears a very large number, the ratio of the electrostatic and gravitational attraction between an electron and a proton. Later this discovery led Eddington to highly curious speculations — totally strange to real physics — concerning the exact value of this and other numbers.

I had long known about these relations, but my real interest in them only arose when I read Bondi's excellent book on cosmology, where he devotes a chapter to them. Since I was looking for a natural explanation for them, I came to think about something that Gamow had told me at our encounter in 1950, namely that the Milky Way, if it were dissolved into protons and electrons, would be at the limit of transparency for ordinary light. This made me investigate what dimensions and what mass a model for a large bounded system of this kind would have, if it simultaneously were at the limit of transparency and close to the above-mentioned density limit. It appeared that this just corresponded to the Eddington relations, but that this density was somewhat larger than the present estimates of the density of the Milky Way system. This result made me tentatively make the following hypothesis concerning the approximate evolution of the system: that it had started as a very thin and large cloud of hydrogen, that subsequently radiation and ionization appeared through atomic collisions, when the density increased, that the radiation by scattering on the electrons slowed down the expansion, which gradually proceeded to the present expansion, all this because the opacity at some stage of the process became large enough. Naturally, this is a most schematic account of the content of my first published, very cautious version; after a talk at an astrophysics conference in Liège in September 1953 (published 1954 in the conference proceedings).

Further investigation of this was slow, partly due to the difficulty of the problem, partly because I was simultaneously heavily occupied with other work. However, in the meantime I got some opportunities to present my views, among others in a talk at a Solvay conference in 1958, published that year.

Only gradually did I get enough confidence in these ideas to draw the in fact obvious conclusion that such a large system as the one in question had to have started out with practically as much anti-hydrogen as ordinary hydrogen — any difference to speak of would be utterly improbable. Thus, some natural process of separation must have taken place, as at least our solar system does not contain more than extremely small amounts of antimatter. The fact that we cannot see whether other stars consist of ordinary matter or antimatter was certainly already realized by many physicists shortly after the discovery of the anti-electron and Dirac's theory of electrons and anti-electrons, although it would take a long time before real antimatter was discovered. I had written about this myself in the book *Cause and Effect* (*Natur och Kultur*, 1935).

I now asked Alfvén — because of his great ingenuity and knowledge of electromagnetic processes in a plasma — whether he could imagine any such separation processes. This was in the spring of 1961, and already in the summer the same year, he could report some promising suggestions to me. He was completely taken by the thought that the annihilation of matter and antimatter not only would be able to contribute to the understanding of the development of the meta-galaxy but also constitute a hitherto overlooked energy source for cosmic processes that were not understood. Through his enthusiasm he contributed considerably to the interest in these problems. A valuable result of that was that Bertel Laurent of Stockholm University, one of my best former students, came to devote himself with fervour to a more thorough study of the development of the meta-galaxy based on the views mentioned, in particular a more realistic treatment of the importance of the annihilation than that of Alfvén and myself, published in connection with his separation attempt (in *Arkiv för fysik*, 1962).

In connection with the problems of cosmology, I have first of all occupied myself with two things after the introductory works: one attempt to show that the so-called fireball radiation — the *pièce de résistance* of cosmologists — gets its natural explanation in the framework of the evolution of the meta-galaxy. And further a careful analysis of Einstein's road to his general theory of relativity, especially his relationship to Mach's idea concerning the forces of inertia.

Einstein's earlier contributions to what he later called the general theory of relativity consisted mainly of some thought experiments, as interesting as they were simple, where he started from the long-known but isolated fact that all bodies get the same acceleration in a given gravitational field. In the most important of the experiments he imagined what would happen in a freely falling room, namely that all the usual effects of gravity disappear. Nowadays, this indeed belongs to the less pleasant experiences of our space travellers. However, already Newton, had he had any reason to reflect upon it, would immediately have known that it is like that. But the novelty was that Einstein traced therein a then unknown principle, which immediately led him to new predictions concerning the behavior of light in a gravitational field, which were demonstrated later. In addition, he realised, as was shown by another of his thought experiments, that the inertial force which appears if the room is instead accelerated in a region without gravitation, is not in any essential way different from what we otherwise call gravitation or force of gravity. Now it is easy for us to see how this consistently led to the entire foundation of the general theory of gravitation of Einstein.

But this is wisdom after the event. In reality this caused him several years of effort, and in the meantime the matter was complicated to him through musings about why nature seemed to prefer uniform motion, why relativity would not be valid for arbitrary motion. Of course it is, but the question is: with respect to what? To Mach, whom Einstein followed regarding this question, it was simple; to him vacuum was empty and and the opposite of matter. Therefore he, unjustly, criticized the absolute space of Newton and had the opinion that the effects of accelerated motion — the forces of inertia — were a kind of gravitation unknown until then, whose source in this case consisted of the collection of celestial bodies in the universe, but which always arose when a body moved nonuniformly with respect to another body, although it takes a very large mass before it is noticeable.

It is understandable that Einstein caught this idea with fervour, which seemed to give him *the general relativity* and thus the main source of inspiration for his cosmology (which in turn inspired the later somewhat more realistic cosmology attempts). However, I cannot see otherwise than that we are here dealing with a common mistake — the power of word over thought — in this case the world vacuum = absolute empty space. In contrast to this the starting point of the principle of equivalence, as we have seen, is a very general *experience* — that all bodies obtain the same acceleration in a gravitational field. But experiences are always limited. Therefore it takes an unusual sense of tact — and in addition certainly luck — to build on such a foundation a new theory, correct in all essentiality. And here the need for explanation is likely to cause trouble, as the explanation, when a really new principle is concerned, has to work with perceptions that lack the generality one had the reason to think. This was true for Newton's theory of gravitation as well as Einstein's theory of relativity, where he himself had been forced to refrain from the kind of explanations that he now demanded concerning his own principle of equivalence.

Now a few words about this. There the concept of inertial frame plays a fundamental role. It is a frame of reference, where gravity is eliminated in the same way as in Einstein's experiment with the freely falling room. But to achieve such weightlessness during a finite time obviously a constant field of gravitation is needed. In a gravitational field of a general kind, i.e. one where the strength and direction of gravitation is changing from point to point in both space and time, one defines a local inertial frame at each point and instant by a transition to the infinitely small, i.e. to let the region surrounding the point and the duration of the observation become smaller and smaller, thus a similar procedure to that by which a tangent surface is defined at every point on a curved surface. In the same way as the geometry of the curved surface is defined by the plane geometry, namely by lengths and directions within an infinitely small region, the principle of equivalence implies that the known physical laws, valid in an inertial frame, first of all the theory of relativity, can be translated to the corresponding laws in a given gravitational field. Thus, for example, the equations of motion of a particle in an arbitrary gravitational field follow from the law of inertia, which is indeed valid in an inertial frame.

An important consequence of this is that space and time in the gravitational field in principle are based on measurements in the local inertial frames, something that Einstein himself has emphasized, in that he at the same time pointed out the analogy with Gauss's theory for the curved surfaces. From this it now also follows that this is valid for the properties of all kinds of isolated systems, be it ordinary bodies or atoms or systems of stars, and in particular for the mass of a given body. It was therefore curious that Einstein overlooked this, in that he, in connection with Mach's idea, thought himself to be able to prove that the mass of a given body increases, when other bodies approach it.

However, there exists something that probably made him doubt somewhat the possibility of unifying the general relativity with the principle of equivalence, namely that his field equations for gravitation, i.e. the analog of Newton's law of gravity, cannot be derived from this principle although the concepts contained in the equations are defined through it. And these equations are independent of coordinates in which they are expressed, which could appear as a verification of the general relativity. However, that this is not the case is clear from the fact that their physical content comes from the inertial frames. That Einstein, with time, got some doubts about the principle of equivalence, but not the general theory of relativity, is indicated by the considerable effort he made to derive the equations of particle motion from the field equations mentioned, while these earlier, also for him, were an obvious and particularly simple result of the equivalence principle.

After this somewhat sketchy account of the various points of view of Einstein towards the theory that he himself had created, certainly something of the most important in the history of physics, I will say something in brief about my attempts to enlarge the principle of equivalence with the help of quantum field theory.

This work, which is far from finished, is based on my ever stronger conviction that gravity is the simplest, the most universal and at the same time the most understood of all forces of nature — a conviction that at present is hardly shared by many physicists. A main reason for this is just the principle of equivalence. But already earlier I found another reason for this, namely that the difficulties of divergencies of quantum electrodynamics probably would disappear if one took gravity into account. That it was possible despite the infinities to enlarge the region of applicability of this theory through the procedure of renormalization — a rational subtraction of the infinite terms — naturally impressed me to a high degree. But I never thought that this was a real solution to the problem.

Before,[2] I mentioned the extremely small length that seemed to play a key role in the five-dimensional theory, which besides the gravitational constant and the velocity of light is formed by the smallest electric charge. But when I got acquainted with the procedure of renormalization, which also affects the latter, I had the opinion that it should be replaced by Planck's quantum of action. By introducing this even somewhat smaller length as one of the three units in the equations of quantum electrodynamics with gravity these become particularly simple. Some time later I found the following expression for the importance of this length for the problem of divergencies. If, namely, any particle whatever could be confined to a sphere of the same order of magnitude as this length, its own gravitation would reach the aforementioned limit for density relative to mass. From this I drew the conclusion that two particles would never be able to come closer together than that, by which the infinities would cease to exist. A remark by Landau in the *Festschrift* on the occasion of Bohr's 70th birthday indicates that he had similar thoughts.

The starting point for the following works, as for the work of 1938 mentioned, is the Dirac equation with gravitation, which had already been developed by several researchers in the beginning of the 1930's — during the period when I had ceased to believe in the importance for quantum field theory of gravitation and its five-dimensional generalization.

Something relatively easy was finding a transformation that disposes of the gravitational field in accordance with the equivalence principle, i.e. at one point with respect to space and time. A considerably more difficult problem, to which I so far have been unable to find a solution, although I think that one should exist, is to derive the field equations for the gravitational field without any additional term for it, by the same method by which the Dirac equation is derived, i.e. through one and the same variational principle. One reason that this should be successful is that the concepts of vacuum and matter have come in a new situation through quantum theory. Instead of being pure opposites they here relate to one another as the normal state of an atom to its excited states. While the vacuum expectation value of a field of force thus vanishes according to quantum theory, this is not valid for the average of the squares, which gives rise to the so-called vacuum fluctuations. Another reason is that here — contrary to what is the case for the electromagnetic field in the usual treatment — one has both the gravitational potentials and their derivatives in the Dirac equation and therefore also in the expression for the variational principle. Thereby one obtains, so to speak, normal field equations also for gravitation. And just via the vacuum fluctuations of the Dirac field these obtain a form which in any case resembles the field equations of Einstein.

Before I end this brief account I will mention one thing which has reinforced my belief that a rational enlargement of the equivalence principle, in connection with the five-dimensional formulation of gravitation and electromagnetism, constitutes a hopeful beginning of a more extensive quantum field theory. It has to do with the unexpected lack of mirror symmetry discovered after a thorough analysis by Lee and Yang, for which they were awarded the Nobel Prize of 1957. The continued investigations led to a new law of nature, which, as Landau showed shortly after the discovery, can be expressed in such a way that the antiparticles are the mirror images of particles. I was namely able to show that this gets a natural interpretation in the framework of the five-dimensional theory (*Nuclear Physics*, 1957).

Interested readers are referred to a review article on this programme (*Nuclear Physics* **B21** (1970)), and a simple derivation of the equivalence principle for the Dirac equation, dedicated to Harald Wergeland on his 60th birthday (*Physica Norvegica* **5**, 1971).

References

1. This article is contained in *The Oskar Klein Memorial Lectures*, Vol. 1, ed. G. Ekspong (World Scientific, 1991), p. 85.
2. The first part of these autobiographical notes can be found in *The Oskar Klein Memorial Lectures*, Vol. 1, ed. G. Ekspong (World Scentific, 1991), p. 103.